The Computation and Theory of Optimal Control

This is Volume 65 in
MATHEMATICS IN SCIENCE AND ENGINEERING
A series of monographs and textbooks
Edited by RICHARD BELLMAN, *University of Southern California*

A complete list of the books in this series appears at the end of this volume.

THE COMPUTATION AND THEORY OF OPTIMAL CONTROL

PETER DYER
Central Instrument Research Laboratory
Imperial Chemical Industries, Ltd.
Nr. Reading, Berkshire, England

STEPHEN R. McREYNOLDS
Jet Propulsion Laboratory
California Institute of Technology
Pasadena, California

1970

ACADEMIC PRESS NEW YORK AND LONDON

ACADEMIC PRESS, INC.
111 Fifth Avenue, New York, New York 10003

United Kingdom Edition published by
ACADEMIC PRESS, INC. (LONDON) LTD.
Berkeley Square House, London W1X 6BA

LIBRARY OF CONGRESS CATALOG CARD NUMBER: 71-91433
AMS 1968 SUBJECT CLASSIFICATION 9340

PRINTED IN THE UNITED STATES OF AMERICA

Contents

PREFACE ix

1. Introduction

1.1 Notation 4

2. Parameter Optimization

2.1 Some Notation and Definitions 8
2.2 Necessary and Sufficient Conditions for a Local Optimum 9
2.3 Numerical Methods 17
Bibliography and Comments 33

3. Optimal Control of Discrete Systems

3.1 Notation 35
3.2 The Problem 37
3.3 Dynamic Programming Solution 37
3.4 Linear Quadratic Problems 42
3.5 Numerical Methods 54
3.6 The Gradient Method 56
3.7 The Newton–Raphson Method 68
3.8 Neighboring Extremal Methods 79
Bibliography and Comments 83

4. Optimization of Continuous Systems

4.1 Notation 85
4.2 The Problem 87
4.3 Dynamic Programming Solution 87
4.4 The Linear Quadratic Problem 96
4.5 Linear Quadratic Problem with Constraints 109
4.6 Stability 114
Bibliography and Comments 115

5. The Gradient Method and the First Variation

5.1 The Gradient Algorithm 117
5.2 The Gradient Algorithm: A Dynamic Programming Approach 118
5.3 Examples 127
5.4 The First Variation: A Stationarity Condition for a Local Optimum 144
Bibliography and Comments 148

6. The Successive Sweep Method and the Second Variation

6.1 Introduction 150
6.2 The Successive Sweep Method 150
6.3 An Alternative Derivation: Control Parameters 154
6.4 Examples 162
6.5 Neighboring Optimal Control 174
6.6 The Second Variation and the Convexity Condition 177
Bibliography and Comments 181

7. Systems with Discontinuities

7.1 Introduction 183
7.2 Discontinuities: Continuous State Variables 184
7.3 Application to Examples 190
7.4 The First Variation: A Stationarity Condition 203
7.5 The Second Variation: A Convexity Condition 204
7.6 Discontinuities in the State Variables 206
7.7 Tests for Optimality 209
7.8 Example 210
Bibliography and Comments 212

8. The Maximum Principle and the Solution of Two-Point Boundary Value Problems

8.1 Introduction 213
8.2 The Maximum Principle 214

8.3 The Linear Quadratic Problem 217
8.4 Techniques for Solving Linear Two-Point Boundary Value Problems 218
8.5 Newton–Raphson Methods for Solving Nonlinear Two-Point Boundary Value Problems 226
8.6 Invariant Imbedding 229
Bibliography and Comments 233

Appendix. **Conjugate Points** 235

INDEX 241

Preface

This book is designed to serve as a textbook at the advanced level for the study of the theory and applications of optimal control. Although a fairly complete treatment of the theory of optimal control is presented, the main emphasis is on the development of numerical algorithms for the computation of solutions to practical problems. Thus this book should be of interest to the practicing systems engineer as well as to the systems student.

The subject matter is developed in order of increasing complexity. First, the fundamental concepts of parameter optimization are introduced, i.e., the idea of the first and second variations and the related numerical algorithms, the gradient, and Newton–Raphson methods. Next, the optimization of multistage systems is considered. Formally the multistage problem can be treated as a parameter optimization problem. However, it is much more convenient to introduce the theory of dynamic programming, which is used as a theoretical basis throughout the remainder of the book. Finally, continuous optimal control problems are treated. Special chapters are devoted to problems with discontinuities and to the solution of two-point boundary value problems.

The main advantage of the dynamic programming approach is that it provides a unified approach to both the theoretical results and the numerical algorithms. The main classical results concerning the first and second variation are established. At the same time the related algorithms are developed.

Many practical examples are used to illustrate the different techniques. Problems are included at the end of each chapter, together with a bibliography of related publications.

Much of the material in the book was drawn from papers written by the authors over the last few years. However, the authors would like to acknowledge that they have been heavily influenced by the works of Bellman and Dreyfus. Also many helpful criticisms have been received from Byron Tapley and Chris Markland. The authors are particularly grateful to Winnie Staniford who typed most of the manuscript.

I

Introduction

The theory of optimization is a subject that has received a great deal of attention from the days of the early Greek mathematicians to the present time. In different ages different facets of the subject have been emphasized, depending upon the influences of prevailing technology. Today, optimization is highly influenced by certain engineering applications, for example, the design of optimal control systems and the computation of optimal trajectories for space vehicles. One distinguishing feature of modern optimization theory is the emphasis on *closed-loop*, as opposed to *open-loop*, solutions. The differences between the two types of solution are best described by means of a simple example. Consider the problem of choosing some parameter α to maximize a function $L(x, \alpha)$, where x is some additional parameter that is specified and will be referred to as the state variable. Clearly the optimal choice of α, α^*, will depend upon the variable x, i.e., $\alpha^* = \alpha^*(x)$. A particular solution $\alpha_1 = \alpha^*(x_1)$ is defined by a specific value of x, x_1. The first form of the solution, $\alpha^*(x)$, is referred to as the *closed-loop* solution, and it specifies the optimal solution for all values of the state variable x. The second form of the solution, α_1, is the open-loop solution, and it specifies the solution for a distinct value of the state variable.

In general the closed-loop solution is more valuable, as it enables a control system to be implemented without prior knowledge of the state variable. However, it is more difficult to obtain, and often an approximation must be used. The emphasis on these solutions has led to the development of two theoretical approaches; the maximum principle (variational approach) and dynamic programming (Hamilton–Jacobi approach).

The dynamic programming approach is particularly appealing, as it can be extended easily to include new classes of optimization problems, especially those in which Markov random processes are present.

The emphasis in the development and application of this modern optimization theory has been influenced not only by the demands of modern technology but also by the products. The most profound effects have been caused by the digital computer. The computer has allowed solutions to be found to many complex problems that are quite intractable analytically. For example, one important application of the computer has been to space travel. Without a computer it is quite unrealistic to attempt to send a man to the moon or to probe space surrounding far-off planets.

The development of better computers has shifted the emphasis from analytical techniques to the development of numerical methods. A large number of algorithms for the solution of optimization problems have appeared over the last few years. Each different technique usually has some desirable feature, and different problems often demand that different techniques be used to obtain their solutions. However, of these various numerical methods, two techniques have been shown to be particularly successful in obtaining answers to a variety of problems: the gradient method and the Newton–Raphson method. From a theoretical point of view these two approaches are interesting as they involve the first and second variations of the function being optimized. Furthermore, the classical necessity and sufficiency conditions for an optimum are obtained as a natural consequence of considering these variations. Although these conditions are primarily of theoretical interest, they must also be thoroughly understood by anyone hoping to implement a numerical algorithm. It is extremely important to be able to distinguish between difficulties that are caused by a poorly formed problem and those that are caused by an error in a computer program.

This book, in both the selection of subject matter and emphasis, reflects the influences just discussed. The basic theory is developed in a unified manner through dynamic programming and the theory of characteristics. Also, in many cases alternative derivations more along the lines of the classical variational approach are presented for comparison. It is hoped that this will prove to be useful to readers whether or not they are familiar with the classical theory. Emphasis is placed on the development of numerical methods for obtaining solutions. In particular, the gradient method and a version of the Newton–Raphson method are treated in some detail for a variety of problems. Techniques for obtaining approximate solutions in the neighborhood of the optimal are also developed.

The main reason for the modern emphasis on computational techniques is that almost all *practical* problems are analytically intractable. Hence, it is appropriate that the majority of the examples considered in this book have no known analytical solution. It should, however, be emphasized that the examples are intended to be tutorial in nature rather than to provide detailed studies of particular problems. Many of the problems would demand a book of this size or more if they were treated exhaustively. Wherever possible references are given so that the interested reader can study the problem in more detail.

The book is divided into eight chapters. The first chapter serves as an introduction to the book and reviews the notation that is used throughout. The second chapter reviews the theory of maxima and minima for a function of several variables. Several numerical techniques for the computation of extremal solutions are discussed, and examples illustrate the properties of the various algorithms. Many of the concepts that are to be used throughout the book are introduced.

The third chapter considers the optimization of multistage decision processes. The theory of dynamic programming is developed and applied to the multistage problem. The solution of the multistage optimization problem is shown to depend upon the solution of a set of backwards difference equations. Although in general these difference equations do not allow an exact solution, such a solution does exist if the system equations are linear and if the performance criterion is quadratic. This solution is derived, in Section 3.4, and its application is illustrated with examples. By using the dynamic programming approach, two numerical algorithms are developed and are shown to be related to the necessary and sufficient conditions for an optimal solution. A technique for generating an approximation to the optimal closed-loop solution in the neighborhood of the optimal is also developed.

In Chapter 4 the theory of dynamic programming is extended to include continuous processes. The dynamic programming solution is shown to be specified in terms of a single partial differential equation. The result is developed in two ways. First, in a formal manner by applying the method of finite differences, and second in a more rigorous manner. Again an exact solution is obtained to the linear quadratic problem, and this solution is applied to several practical examples. Problems with linear terminal constraints are also discussed.

However, in general exact solutions are not possible, and numerical solutions must be found. The next three chapters are devoted to a discussion of such techniques. In Chapter 5 the gradient method for continuous systems is derived and applied to several examples. The stationarity condition for a local optimum is also derived. The successive sweep algorithm, a Newton–Raphson algorithm, is derived and applied to examples in Chapter 6. The second variation is examined, and the necessary and sufficient conditions for a local optimum are completed.

In Chapter 7 the successive sweep algorithm is extended to problems with discontinuities. Again necessary and sufficient conditions are derived.

In Chapter 8 dynamic programming is used to derive the maximum principle. The maximum principle leads to a two-point boundary value problem, and hence the solution of linear and nonlinear two-point boundary value problems is considered. Finally, invariant imbedding is discussed.

A small bibliography of relevant material is included at the end of each chapter. A general listing of supplemental material is presented for the convenience of readers who may wish to examine other approaches. Also included are problems designed to encourage the reader to carry out developments omitted from the text.

There are so many different numerical algorithms and combinations of such algorithms that it is not possible to treat all methods in equal detail. Thus, instead of attempting a "cookbook" of algorithms, the authors have concentrated on presenting in detail a few of the more successful techniques. The authors are not familiar with any problems that cannot be solved with these algorithms. However, the reader should appreciate that these techniques will not instantly solve all problems. A great deal of care must be exercised in the formulation of complex problems; for example, the choice of the correct independent variable is sometimes vital. Time spent in formulating the problem is usually very rewarding and can save vast amounts of computer time. There is still no substitute for physical intuition and analytical skills.

1.1 NOTATION

Unfortunately, difficulties arise in optimization notation, and different authors use different notations. Here, *vector notation* will be used. Almost all the quantities are vectors or matrices. Vectors are usually denoted by lowercase letters, matrices by capitals. Unless otherwise specified, a vector is understood to be a column vector. Superscripts and subscripts $i, j, k, \ldots, 1, 2, 3, \ldots$, unless otherwise indicated, denote a particular component of a vector or matrix. No basic distinction is employed between superscripts and subscripts, for example,

$$x = \begin{bmatrix} x_1 \\ x_2 \\ \vdots \\ x_n \end{bmatrix} = \begin{bmatrix} x^1 \\ x^2 \\ \vdots \\ x^n \end{bmatrix}, \qquad A = \begin{bmatrix} A_{11} & A_{12} & & A_{nn} \\ A_{21} & & & \\ A_{n1} & & \cdots & A_{nn} \end{bmatrix}$$

The product of two matrices or of a matrix and a vector is denoted by writing the symbols next to each other. As a rule, no subscripts or summation signs are used, for example, A times x: Ax. Superscript T denotes the transpose, for example,

$$x^{\mathrm{T}} = [x_1, x_2, \ldots, x_n], \qquad A^{\mathrm{T}} = \begin{bmatrix} A_{11} & A_{21} & \cdots & A_{n1} \\ \vdots & & & \\ A_{1n} & & \cdots & A_{nn} \end{bmatrix}$$

Partial derivatives are usually denoted by subscripts. For example, if $f(x)$ is a scalar function of the vector x, then

$$\partial f/\partial x = [\partial f/\partial x_1 \cdots \partial f/\partial x_n]$$

is written f_x. If f were a vector function, then

$$f_x = \begin{bmatrix} f^1_{x_1} & f^1_{x_2} & f^1_{x_3} & \cdots \\ \vdots & & & \\ f^n_{x_1} & & & \end{bmatrix}$$

The superscripts that were used in writing $f^i_{x_j}$ are for notational convenience.

In some cases subscripted variables that do not denote partial derivatives are defined in the text. In such cases the subscript is used as a mnemonic. For example, the quadratic function $L(x)$ will be written

$$L(x) = L_x x + \tfrac{1}{2} x^{\mathrm{T}} L_{xx} x$$

although L_x denotes $\partial L/\partial x$ only when $x = 0$. The subscripts on the variable Z are *always* mnemonics. If there are two subscripts, transposing the subscripts is equivalent to transposing the matrix, for example, $L_{ux} = L_{xu}^{\mathrm{T}}$.

Products of the form $V_x f_{xx}$ are always given by

$$V_x f_{xx} = \sum_i V_{x_i} f^i_{x_j x_k}$$

Functions are usually functions of several variables, often three or four. In order to reduce the complexity of the expressions, arguments are often either omitted or condensed. Frequently the arguments are themselves functions of the independent variable time t. In such cases these arguments may be replaced by the single argument t, for example,

$$f(x(t), u(t), t) = f(t)$$

In multistage problems the analog to the time t is the stage, k, i.e.,

$$h(x(k), u(k), k) = h(k)$$

The Euclidean length of a vector is indicated by placing it between the symbol $\| \; \|$, for example,

$$\left(\sum_{i=1}^{n} x_i^2\right)^{1/2} = \| x \| = (x^{\mathrm{T}}x)^{1/2}$$

If A is a matrix, then the scalar $(x^{\mathrm{T}}Ax)^{1/2}$ may be written

$$(x^{\mathrm{T}}Ax)^{1/2} = \| x \|_A$$

If A is a symmetric matrix then $A > 0$ indicates A is positive definite, and $A < 0$ indicates A is negative definite. Some nonstandard functions are also used:

arg: obtain the argument of
ext: extremize (maximize or minimize)
max: maximize
min: minimize
sgn: the sign of

For example,

$$\alpha^* = \arg \max_{\alpha} \{f(\alpha)\}: \quad \text{set } \alpha^* \text{ to be the value of } \alpha \text{ that maximized } f(\alpha)$$

The overdot is used to denote the total derivative of a function with respect to the independent variable:

$$\dot{f}(x(t), u(t), t) = f_x \dot{x} + f_u \dot{u} + f_t$$

In Chapter 3 square brackets are used to denote a sequence, for example,

$$x[i, j] = \{x(i), x(i + 1), \ldots, x(j)\}$$

2

Parameter Optimization

In this chapter the optimization of a function of several variables is reviewed. Only maximization problems will be considered because any minimization problem can be reformulated as a maximization problem by changing the sign in front of the function that is to be optimized. First, definitions and notation will be decided upon so that certain concepts, to be used throughout the book, may be introduced. The function that is to be optimized is referred to as the *performance index*, and its arguments are classified as either *control parameters* or *state variables*. Control parameters are the variables that must be chosen optimally. State variables are other quantities which, although not chosen explicitly, affect the performance index. The introduction of state variable notation allows the concepts of open-loop and closed-loop solutions to be introduced. An understanding of these particular ideas becomes increasingly important in future chapters.

Some mathematical theory relating to local maxima is developed. This theory is derived by examining the first and second variations of the performance index. Two necessary conditions for a local maxima are obtained. The first, the *stationarity* condition, results from an examination of the first variation. The other, the *convexity* condition, follows from considering the second variation. A sufficiency theorem is obtained from the stationarity condition and a strengthened convexity condition, after which some simple examples illustrate this theory. Next, some numerical methods are discussed, and the gradient and the Newton–Raphson methods are developed in some detail. It is these two methods that form the basis for the numerical techniques that are discussed in later chapters. Some examples are included to illustrate the properties of both methods.

2.1 SOME NOTATION AND DEFINITIONS

Two types of variables are used. The first, designated *control. parameters*, are denoted by a p-dimensional vector, α,

$$\alpha = \begin{bmatrix} \alpha_1 \\ \alpha_2 \\ \vdots \\ \alpha_p \end{bmatrix} \tag{2.1}$$

The other variables are designated *state variables* and are denoted by an n-dimensional vector, x, i.e.,

$$x = \begin{bmatrix} x_1 \\ x_2 \\ \vdots \\ x_n \end{bmatrix} \tag{2.2}$$

The value of the performance index is denoted by J,

$$J = L(x, \alpha) \tag{2.3}$$

where L is a scalar single-valued function of the variables x and α. If the variables x are known, the problem is said to be *deterministic* and involves the choice of α to maximize the return J.

The solutions to these problems take one of two forms. One, the *open-loop* solution, is given in terms of specific numerical values for α. In the other, the *closed-loop* solution, α is specified as a function of the state variables. The function specifying the control in terms of the state, denoted by $\alpha^{\text{opt}}(x)$, is referred to as the *optimal control law*. The closed-loop solution is more general, more desirable, and generally more difficult to find. If the optimal control law is substituted back into the performance index, a function of the state is obtained. This function, which is denoted by $V^{\text{opt}}(x)$, is referred to as the *optimal return function*,

$$V^{\text{opt}}(x) = L(x, \alpha^{\text{opt}}(x)) \tag{2.4}$$

Constraints are divided into two kinds: *equality constraints*, which have the form,

$$M(x, \alpha) = 0 \tag{2.5}$$

and *inequality constraints*, which have the form

$$M(x, \alpha) \leqslant 0 \tag{2.6}$$

where M may be a scalar or vector function.

Sometimes, especially in obtaining closed-loop solutions to problems with constraints, it may be more convenient to rewrite these constraints as

$$M(x, \alpha) = K \tag{2.7a}$$

or

$$M(x, \alpha) \leqslant K \tag{2.7b}$$

where K is some specified constant or vector, and is referred to as the *constraint level.* The closed-loop solution may also be a function of the constraint level K.

A choice of α that satisfies the constraints is said to be *admissible.* When the constraints are present the problem is to find the admissible value of the control parameters that optimizes the performance index.

2.2 NECESSARY AND SUFFICIENT CONDITIONS FOR A LOCAL OPTIMUM

Next some mathematical theory concerning locally optimal solutions will be developed. First a local maximum of the performance index L with respect to α is defined. Let α^* denote a local maximum for a deterministic problem. Then there exists a positive number e such that

$$L(x, \alpha) \leqslant L(x, \alpha^*) \tag{2.8}$$

for all admissible α such that $| \alpha - \alpha^* | < e$.

In other words, α^* is a local maximum for L if $L(x, \alpha^*)$ is greater than or equal to $L(x, \alpha)$ for all admissible α in a neighborhood of α^*.

Local optimal solutions are of interest because of the existence of a body of mathematical theory that may be used to determine them. Clearly, the definition given is of little *practical* use in determining an optimal solution, for it implies that the return L must be evaluated for an infinite number of admissible values of α before choosing the optimal parameters α^*. However, provided L has continuous derivatives with respect to α, more useful results may be obtained by considering the following Taylor series expansion of the function $L(x, \alpha)$ around a local maximum α^*:

$$L(x, \alpha) = L(x, \alpha^*) + L_\alpha(x, \alpha^*)\, \delta\alpha + \tfrac{1}{2}\, \delta\alpha^T L_{\alpha\alpha}\, \delta\alpha + O(\delta\alpha^3) \qquad (2.9)^\dagger$$

where $\delta\alpha = \alpha - \alpha^*$. Defining $\delta J = L(x, \alpha) - L(x, \alpha^*)$ and rewriting this equation gives

$$\delta J = L_\alpha(x, \alpha^*)\, \delta\alpha + \tfrac{1}{2}\, \delta\alpha^T L_{\alpha\alpha}\, \delta\alpha + O(\delta\alpha^3) \tag{2.10}$$

† Subscripts here are used to denote partial derivatives, i.e., $\partial L/\partial\alpha = L_\alpha$.

The magnitude of the remainder term $O(\delta\alpha^3)$ will depend upon the properties of the partial derivatives of L with respect to u. (In the previous equation it is assumed that the third partial derivatives are continuous and bounded.)

The first necessary condition concerns the first term, $L_\alpha\, \delta\alpha$, in the foregoing expansion. This first term is referred to as the *first variation*, and is denoted by $\delta^1 J$, i.e., $\delta^1 J = L_\alpha\, \delta\alpha$.

One obvious property that the local maximum must exhibit, for problems without constraints, is that

$$L_\alpha(x, \alpha^*) \equiv 0 \tag{2.11}$$

This is referred to as the *stationarity* condition. That this condition is necessary may easily be shown by contradiction. Consider the control α defined by

$$\alpha = \alpha^* + \tilde{\epsilon} L_\alpha{}^{\mathrm{T}} \qquad (2.12)^\dagger$$

where $\tilde{\epsilon} > 0$ can be arbitrarily small. Substituting $\delta\alpha = \alpha - \alpha^* = \tilde{\epsilon} L_\alpha{}^{\mathrm{T}}$ into the foregoing expansion of L, Equation (2.10), gives

$$\delta J = \tilde{\epsilon} \, \| L_\alpha \|^2 + \tfrac{1}{2}\tilde{\epsilon}^2 L_\alpha L_{\alpha\alpha} L_\alpha{}^{\mathrm{T}} + O(\tilde{\epsilon}^3) \tag{2.13}$$

Now provided L_α is nonzero, we may choose $\tilde{\epsilon}$ so small that the first term dominates the expansion and such that $\tilde{\epsilon} \, \| L_\alpha \| < \epsilon$ for some arbitrary positive ϵ. Thus, in any neighborhood containing α^* there exists an α such that $\delta J > 0$, or $L(x, \alpha) > L(x, \alpha^*)$. But this contradicts the original hypothesis that α^* is a local maximum, hence, Equation (2.11) must be true.

Now if $L_\alpha(x, \alpha^*) \equiv 0$, the first term in the expansion of $L_\alpha(x, \alpha)$, Equation (2.10), vanishes, leaving

$$\delta J = \tfrac{1}{2}\, \delta\alpha^{\mathrm{T}} L_{\alpha\alpha}\, \delta\alpha + O(\delta x^3) \tag{2.14}$$

The leading term in this expansion is known as the *second variation* and is denoted by $\delta^2 J$, i.e., $\delta^2 J = \frac{1}{2}\, \delta\alpha^{\mathrm{T}} L_{\alpha\alpha}\, \delta\alpha$. Clearly for δJ to be nonpositive $[L(x, \alpha) \leqslant L(x, \alpha^*)]$,

$$\delta^2 J = \delta\alpha^{\mathrm{T}} L_{\alpha\alpha}\, \delta\alpha \leqslant 0 \tag{2.15}$$

for all admissible values of $\delta\alpha$. In other words, the eigenvalues of the real symmetric matrix $L_{\alpha\alpha}$ must be nonpositive. This is called the *convexity condition*. Suppose there existed a vector v such that $v^{\mathrm{T}} L_{\alpha\alpha} v > 0$, a control α could be constructed, such as

$$\alpha = \alpha^* + \tilde{\epsilon} v \tag{2.16}$$

† Superscript T is used to denote the transpose.

where $\tilde{\epsilon}$ may be arbitrarily small. Then substituting $\delta\alpha = \alpha - \alpha^* = \tilde{\epsilon}v$ into Equation (2.14) gives

$$\delta J = \tilde{\epsilon}^2 v^{\mathrm{T}} L_{\alpha\alpha} v + O(\tilde{\epsilon}^3) \tag{2.17}$$

Now, provided $L_{\alpha\alpha}$ is nonzero, $\tilde{\epsilon}$ may be chosen such that the first term, $\tilde{\epsilon}^2 v^{\mathrm{T}} L_{\alpha\alpha} v$, dominates the expansion and such that $\tilde{\epsilon}v < \epsilon$ for some arbitrary positive constant ϵ. But for this choice of $\tilde{\epsilon}$ this first term must be positive, i.e.,

$$L(x, \alpha) > L(x, \alpha^*) \tag{2.18}$$

which contradicts the original hypothesis that α^* is a local maximum. Thus, $L_{\alpha\alpha}$ must be negative semidefinite.

Both the stationarity and convexity conditions are necessary conditions for a local maximum of the problem without constraints. It is apparent that similar arguments may be applied to a local minimum.

Next a sufficiency theorem will be established for a local maximum of the deterministic problem.

THEOREM 2.1. Let $L_{\alpha\alpha}(x, \alpha^*)$ exist and be bounded and continuous. Also assume that

$$L_\alpha(x, \alpha^*) = 0 \tag{2.19}$$

and

$$\delta\alpha^{\mathrm{T}} L_{\alpha\alpha}(x, \alpha^*)\, \delta\alpha < 0 \tag{2.20}$$

for arbitrary $\delta\alpha$. Then α^* is a local maximum.

PROOF. The theorem follows by showing that there exists an ϵ such that $|\alpha - \alpha^*| < \epsilon$ implies that $\delta J < 0$, where δJ is defined above Equation (2.10). Since L_α satisfies Equation (2.19), it follows that δJ may be simplified to Equation (2.14). Define λ as the largest eigenvalue of $L_{\alpha\alpha}$. Then λ must be negative [Equation (2.20)]. Hence, for $|\delta u| < \epsilon$,

$$\delta J < \lambda\epsilon^2 + O(\epsilon^3) \tag{2.21}$$

which may be made negative by choosing ϵ so that the term $\lambda\epsilon^2$ dominates the expression.

Note that the conditions equations (2.19) and (2.20) are very similar to the necessary conditions just derived. The first condition is identical to the stationarity condition. The other resembles the convexity condition except that the weak inequality has been strengthened to a strong inequality. The condition given by Equation (2.20) will be referred to as the *strengthened convexity condition.*

If side constraints are present the foregoing results require some modification. The local maximum α^* must also satisfy [cf Equation (2.5)],

$$M(x, \alpha^*) = 0 \tag{2.22}$$

Hence, the variations $\delta\alpha$ may no longer be arbitrary because $\alpha = \alpha^* + \delta\alpha$ must satisfy the side constraints. In other words, $\delta\alpha$ must satisfy an equation of the form

$$0 = M_\alpha(x, \alpha^*)\,\delta\alpha + \tfrac{1}{2}\,\delta\alpha^{\mathrm{T}} M_{\alpha\alpha}(x, \alpha^*)\,\delta\alpha + O(\delta\alpha^3) \tag{2.23}$$

Now let us assume that $\partial M/\partial\alpha_1 = M_{\alpha_1}$ is nonzero. Equation (2.23) implies that

$$\delta\alpha_1 = -M_{\alpha_1}^{-1}[M_{\alpha'}\,\delta\alpha' + \tfrac{1}{2}\,\delta_\alpha{}^{\mathrm{T}} M_{\alpha\alpha}\,\delta_\alpha] \tag{2.24}$$

where α' is the vector $(\alpha_2, \alpha_3, \ldots, \alpha_p)^{\mathrm{T}}$. If $\delta\alpha_1$ is chosen by Equation (2.24), it is clear that the foregoing constraint can be satisfied for arbitrary $\delta\alpha'$. Using this equation to eliminate $\delta\alpha_1$ from Equation (2.10), we obtain

$$\delta J = (L_{\alpha'} - L_{\alpha_1} M_{\alpha_1}^{-1} M_{\alpha'})\,\delta\alpha' + \tfrac{1}{2}\,\delta\alpha^{\mathrm{T}}(L_{\alpha\alpha} - L_{\alpha_1} M_{\alpha_1}^{-1} M_{\alpha\alpha})\,\delta\alpha + O(\delta\alpha^3) \tag{2.25}$$

Since $\delta\alpha'$ is arbitrary, it follows that, for a maximum

$$L_{\alpha'} - L_{\alpha_1} M_{\alpha_1}^{-1} M_{\alpha'} = 0 \tag{2.26}$$

Now define $\nu = -L_{\alpha_1} M_{\alpha_1}^{-1}$. Then this stationarity condition may be restated as follows:

$$L_\alpha + \nu M_\alpha = 0 \tag{2.27}$$

The first equation in Equation (2.27), namely, $L_{\alpha_1} + \nu M_{\alpha_1} = 0$, follows from the definition of ν. The remainder follows from Equation (2.26).

The geometrical interpretation of this condition is quite clear. It implies that the vectors L_α and M_α are parallel. Classically the parameter ν is referred to as a *Lagrange multiplier*.

The convexity condition is obtained as follows. Assuming the stationarity condition [Equation (2.27)] is satisfied, and replacing $L_{\alpha_1} M_{\alpha_1}^{-1}$ by ν, Equation (2.25) for δJ becomes

$$\delta J = \tfrac{1}{2}\,\delta\alpha^{\mathrm{T}}(L_{\alpha\alpha} + \nu M_{\alpha\alpha})\,\delta\alpha + O(\delta\alpha^3) \tag{2.28}$$

Note that $\delta\alpha$, not $\delta\alpha'$, appears in this equation. However, as $\delta\alpha$ is required to satisfy Equation (2.23), Equation (2.24) must hold. Again Equation (2.24) may be used to eliminate $\delta\alpha_1$, thus obtaining a second-order expansion of L in terms of $\delta\alpha'$. Note that to do this only the first-order term in the expansion equation (2.23) is needed to form $\delta\alpha'$. Hence, the convexity condition is simply that

$$\delta\alpha^{\mathrm{T}}(L_{\alpha\alpha} + \nu M_{\alpha\alpha})\, \delta\alpha \leqslant 0 \tag{2.29}$$

for all $\delta\alpha$ such that $M_\alpha\, \delta\alpha = 0$. In other words, $v^{\mathrm{T}}(L_{\alpha\alpha} + \nu M_{\alpha\alpha})\, v$ must be nonpositive for all vectors v perpendicular to M_α.

If inequality constraints, such as those given by Equation (2.6), are present then two cases may occur. Either the constraint $M < 0$ or $M = 0$ must hold. In the former case, it is clear that the constraint does not affect admissible variations about the maximum, and hence it can be ignored when the conditions for the local maximum are specified. However, in the latter case, $M = 0$, the stationarity and convexity conditions of Equations (2.27) and (2.29) must be satisfied. In addition, variations of the form

$$\delta\alpha = -\epsilon M_\alpha(x, \alpha^*), \qquad \epsilon > 0 \tag{2.30}$$

may be considered. These variations are such that $M(x, \alpha) < 0$ were excluded when the equality constraint was discussed. Substituting this value for $\delta\alpha$ into the expansion for δJ, Equation (2.10) gives

$$\delta J = -\epsilon L_\alpha M_\alpha{}^{\mathrm{T}} + O(\epsilon^2) \tag{2.31}$$

Now, choosing ϵ such that the first term dominates this equation leads to the condition

$$-\epsilon L_\alpha M_\alpha{}^{\mathrm{T}} \leqslant 0 \tag{2.32}$$

for a local maximum. But since the stationarity condition, Equation (2.27), applies, the equation may be written

$$\epsilon\nu \, \| M_\alpha \|^2 \leqslant 0 \tag{2.33}$$

Hence,

$$\nu \leqslant 0 \tag{2.34}$$

This is referred to as the *sign of the multiplier rule.*

2.2.1 Interpretation of the Lagrange Multiplier

An interpretation of the Lagrange multiplier may be obtained by considering the effects of small changes in the constraint level K. In the foregoing analysis K was taken to be zero. First the constraints are adjoined to the index to form an augmented performance index J^*,

$$J = L(x, \alpha) + \nu^{\mathrm{T}}(M(x, \alpha) - K) \tag{2.35}$$

where ν is the Lagrange multiplier. The solution, in particular the value of the optimal return $J^*(x, \alpha^{\text{opt}}(x))$, will depend on the value of K. The Lagrange multiplier is interpreted as

$$\nu = -dJ^{*\text{opt}}/dK \tag{2.36}$$

i.e., ν is equivalent to the total rate of change of the optimal value of the performance index for a small change in the constraint level K.

This assertion follows immediately from a consideration of the first variation of the modified performance index, Equation (2.35),

$$dJ^{*\text{opt}} = (L_\alpha + \nu^{\text{T}} M_\alpha)\, d\alpha + (M - K)^{\text{T}}\, d\nu - \nu^{\text{T}}\, dK$$

But at the optimum the first two terms vanish. Hence, $dJ^{*\text{opt}} = -\nu^{\text{T}}\, dK$ which is the desired result.

Note that the introduction of the Lagrange multiplier allows the original problem to be replaced by the problem of extremizing an augmented performance index J^* where

$$J^* = L(x, \alpha) + \nu M(x, \alpha)$$

In terms of this problem, the stationarity and convexity conditions are

$$\partial J^*/\partial\alpha = \partial J^*/\partial\nu = 0$$

and $v^{\text{T}}(\partial^2 J^*/\partial\alpha^2)\, v < 0$ for all vectors v such that $M_\alpha v = 0$. The approach of forming an augmented performance index in this fashion will be used throughout the book.

2.2.2 Abnormality

If, at the optimum, either

$$M_\alpha^{\,i} = 0, \qquad i \in [1, 2, \ldots, p] \tag{2.37}$$

or there exists a vector λ such that

$$\lambda^{\text{T}} M_\alpha = 0 \tag{2.38}$$

the system is said to be *abnormal.* The first equation (2.37) implies that in the neighborhood of the optimum, to first order, the ith constraint has no effect on the solution. The second equation (2.38) implies that (again to first order, in the neighborhood of the optimum) the constraints are *not* independent. In such cases it is clear that the necessary conditions just derived do not necessarily hold. Hence, abnormal cases are excluded implicitly from this analysis.

2.2.3 Examples

EXAMPLE 2.1. Consider the problem of maximizing the performance index J where

$$J = L_x x + L_\alpha \alpha + \tfrac{1}{2} L_{xx} x^2 + L_{\alpha x} x \alpha + \tfrac{1}{2} L_{\alpha\alpha} \alpha^2 \tag{2.39}$$

Here x and α are scalar variables. In this example the coefficients L_x, L_α, etc., are *not* defined as partial derivatives. (They are, however, equivalent to the partial derivatives of J evaluated at $x = 0$ and $\alpha = 0$.) The subscripts are merely retained for notational convenience. The stationarity condition is $\partial J/\partial\alpha = 0 = L_\alpha + L_{\alpha x} x + L_{\alpha\alpha}\alpha$. Solving this equation for α gives the optimal control law, viz,

$$\alpha^{\mathrm{opt}}(x) = -L_{\alpha\alpha}^{-1}(L_\alpha + L_{\alpha x} x) \tag{2.40}$$

Next, the convexity condition is examined to determine whether or not the solution is a maximum. The solution is a maximum if $\partial^2 J/\partial\alpha^2 < 0$ or $L_{\alpha\alpha} < 0$. There is a local minimum if $L_{\alpha\alpha} > 0$.

Substituting the optimal value of α, Equation (2.40), back into the performance index, Equation (2.39) gives the optimal value of the return V^{opt}

$$V^{\mathrm{opt}}(x) = -\tfrac{1}{2} L_\alpha L_{\alpha\alpha}^{1} L_\alpha + (L_x - L_\alpha L_{\alpha\alpha}^{-1} L_{\alpha x})\, x + \tfrac{1}{2}(L_{xx} - L_{\alpha x}^2 L_{\alpha\alpha}^{-1}) x^2$$

EXAMPLE 2.2. Consider a foil suspended in a moving gas (for example, the wing of an airplane). What angle of attack α should be chosen to

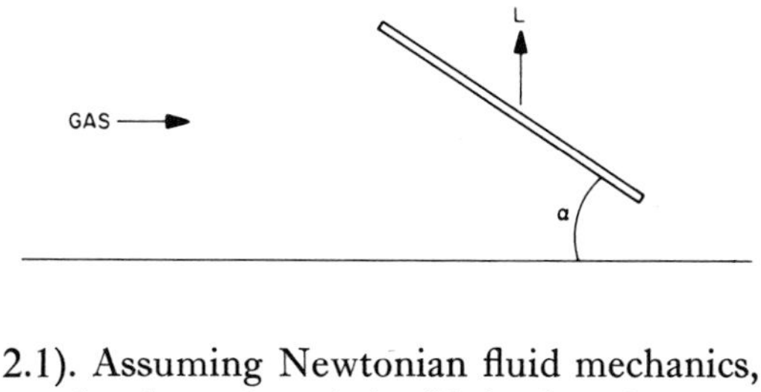

FIG. 2.1. A foil suspended in a moving gas (Example 2.2).

maximize the lift L (Figure 2.1). Assuming Newtonian fluid mechanics, the relationship between the angle of attack and the lift is given by

$$L = K \sin^2 \alpha \cos \alpha, \qquad 0 < \alpha < \pi$$

where the positive parameter K depends upon the physical properties of the gas, the foil, etc. Applying the stationarity condition

$$\partial L/\partial\alpha = 2K \sin\alpha \cos^2\alpha - K \sin^3 \alpha = 0$$

The admissible value of α that satisfies this equation is given by

$$\alpha = \tan^{-1} 2 = 54°44'$$

Examining the convexity condition for $\alpha = 54°44'$,

$$\partial^2 L/\partial\alpha^2 = 2K \cos^3 \alpha - 7K \sin^2 \alpha \cos \alpha \simeq -2.3K$$

which confirms that this value of α corresponds to a maximum.

EXAMPLE 2.3. Consider the problem of choosing the control parameters α_1 and α_2 to maximize the performance index J,

$$J = 1 - \alpha_1{}^2 - \alpha_2{}^2$$

subject to the constraint M

$$M = \alpha_1 + \alpha_2 - 4 = 0$$

The augmented performance index J^* is

$$J^* = 1 - \alpha_1{}^2 - \alpha_2{}^2 + \nu(\alpha_1 + \alpha_2 - 4)$$

The stationarity condition is obtained by partially differentiating J^* with respect to $\alpha = [\alpha_1, \alpha_2]$, i.e.,

$$\partial J^*/\partial\alpha = [-2\alpha_1 + \nu, -2\alpha_2 + \nu] = [0, 0]$$

Hence, it is obvious that

$$\alpha_1 = \alpha_2 = \nu/2 \tag{2.41}$$

Also, the derivative of J^* with respect to ν must be zero, i.e.,

$$\partial J^*/\partial\nu = \alpha_1 + \alpha_2 - 4 = 0 \tag{2.42}$$

Substituting for α_1 and α_2 from Equations (2.41) and (2.42) gives $\nu = 4$ and $\alpha_1 = \alpha_2 = 2$.

Next, the convexity condition must be checked. The condition is that $\delta\alpha^T J^*_{\alpha\alpha}\, \delta\alpha \leqslant 0$ for all $\delta\alpha$ such that $M_\alpha\, \delta\alpha = 0$. Now the condition $M_\alpha\, \delta\alpha = 0$ implies that

$$[1 \quad 1]\begin{bmatrix} \delta\alpha_1 \\ \delta\alpha_2 \end{bmatrix} = \delta\alpha_1 + \delta\alpha_2 = 0$$

or $\delta\alpha_2 = -\delta\alpha_1$.

This relation may be used to eliminate $\delta\alpha_2$ from the second variation and so the convexity condition becomes

$$[\delta\alpha_1, -\delta\alpha_1]\begin{bmatrix} -2 & 0 \\ 0 & -2 \end{bmatrix}\begin{bmatrix} \delta\alpha_1 \\ -\delta\alpha_1 \end{bmatrix} \leqslant 0 \qquad \text{for all } \delta\alpha_1$$

The left-hand side is equal to $-4\,\delta\alpha_1{}^2$ and hence the condition is satisfied.

2.3 NUMERICAL METHODS

In general it will not be possible to find the optimal control parameter α analytically, especially if α is a multidimensional vector. Hence, some form of numerical algorithm must be applied to find the optimum. Two basic algorithms, the gradient and the Newton–Raphson, will be discussed in detail. It is these algorithms that form the basis for the dynamic optimization procedures in the following chapters. The application of the algorithms will be illustrated with several examples. Finally, some of the other numerical techniques will be discussed.

2.3.1 The Gradient Algorithm

The gradient method is motivated from a consideration of a first-order expansion of the performance index about some nominal J^k, viz,

$$J^{k+1}(x, \alpha^{k+1}) = J^k(x, \alpha^k) + L_\alpha(x, \alpha^k)\, \delta\alpha + O(\delta\alpha^2) \tag{2.43}$$

where the superscript k refers to the iteration and $\delta\alpha = \alpha^{k+1} - \alpha^k$ is chosen such that the term $O(\delta\alpha^2)$ is small when compared with $L_\alpha(x, \alpha^k)\, \delta\alpha$. Now, as was shown in the derivation of the stationarity condition, if $\delta\alpha$ is chosen by

$$\delta\alpha = \epsilon L_\alpha^{\mathrm{T}} \tag{2.44}$$

J^{k+1} will be greater than J^k.

Thus the gradient algorithm generates a sequence of nominal solutions $\alpha^0, \alpha^1, \ldots, \alpha^k$ that hopefully converges to the optimal solution by the following formula

$$\alpha^{k+1} = \alpha^k + \epsilon L_\alpha^{\mathrm{T}}(x, \alpha^k) \tag{2.45}$$

A geometrical interpretation of the gradient algorithm is given in Figure 2.2 for a two-dimensional example. The parameter ϵ, a positive scalar, is chosen to ensure the convergence of the solutions. Now, as just shown, this choice of α^{k+1} only ensures that $L^{k+1} > L^k$. To show that α^k will converge to the optimal value α^{opt}, consider a vector, e^{k+1}, the error in the control parameter α^{k+1},

$$e^{k+1} = \alpha^{k+1} - \alpha^{\mathrm{opt}} \tag{2.46}$$

and define the vector function F where

$$F(\alpha) = \alpha + \epsilon L_\alpha(\alpha) \tag{2.47}$$

Thus,

$$\alpha^{k+1} = F(\alpha^k), \qquad \alpha^{\text{opt}} = F(\alpha^{\text{opt}}) \tag{2.48}$$

Thus, Equations (2.47) and (2.48) may be used to rewrite Equation (2.46)

$$e^{k+1} = F(\alpha^k) - F(\alpha^{\text{opt}}) \tag{2.49}$$

Now, assuming that α^k is near to α^{opt}, $F(\alpha^k)$ may be expanded in a Taylor series about $F(\alpha^{\text{opt}})$, giving, to first order,

$$e^{k+1} = F(\alpha^{\text{opt}}) + F_\alpha(\alpha^{\text{opt}})(\alpha^k - \alpha^{\text{opt}}) - F(\alpha^{\text{opt}}) + O(\alpha^k - \alpha^{\text{opt}})^2 \tag{2.50}$$

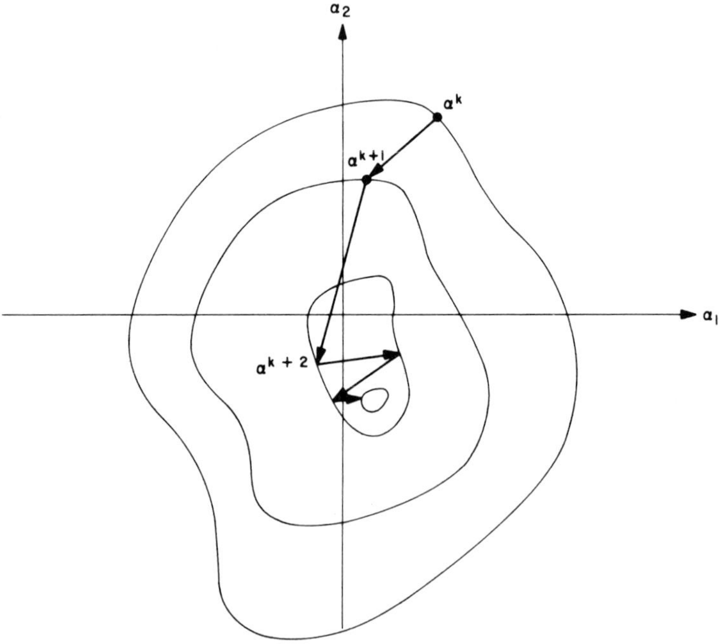

FIG. 2.2. The gradient algorithm.

Noting that by definition $\alpha^k - \alpha^{\text{opt}} = e^k$ and collecting terms leads to

$$e^{k+1} = F_\alpha e^k + O(e^k)^2 \tag{2.51}$$

Here $F_\alpha(\alpha^{\text{opt}})$ is given by [cf Equation (2.47)]

$$F_\alpha = I + \epsilon L_{\alpha\alpha}(\alpha^{\text{opt}}) \tag{2.52}^\dagger$$

† It is tacitly assumed that $L_{\alpha\alpha}$ exists and is continuous in the neighborhood of α^{opt}.

where I is the unit matrix. Convergence will result, i.e., $e^k \to 0$, provided the eigenvalues of F_α are less than unity in absolute value, i.e.,

$$|I + \epsilon L_{\alpha\alpha}(\alpha^{\text{opt}})| < 1$$

This is possible since $L_{\alpha\alpha}$ is a negative definite matrix in the neighborhood of the maximum (the convexity condition). In these circumstances e^{k+1} is proportional to e^k and the technique is said to have first-order convergence.

If side constraints are present, the basic procedure must be amended. There are several alternative approaches, although only two will be discussed here—the *penalty function technique* and the *gradient projection technique.*

The *penalty function technique* is particularly useful if the side constraints need only be satisfied approximately, which is often the case in practical situations. A strictly convex function of the side constraints is adjoined to the performance index to form an augmented index J^*,

$$J^* = J - c(M - K)^{2r}$$

Here, $c > 0$ is a weighting parameter and r is an integer, often unity. Now the normal gradient algorithm may be used to maximize J^*. The weighting parameter c is chosen to ensure that the side constraint is met to the desired accuracy. Increasing c will tend to reduce the error in the side constraint.

The second technique involves modifying the basic algorithm. Again an augmented index is formed except that here the constraint itself is adjoined to the index, i.e.,

$$J^* = J + (\nu^k)^{\mathrm{T}} (M - K)$$

where ν^k is a vector of multipliers. The reader can easily verify that the gradient correction to a nominal solution α^k becomes

$$\delta\alpha = \alpha^{k+1} - \alpha^k = \epsilon(L_\alpha{}^{\mathrm{T}} + \nu^k M_\alpha{}^{\mathrm{T}})|_{\alpha=\alpha^k} \tag{2.53}$$

All that remains is to determine the vector ν^k. This is done by choosing ν^k to ensure a predetermined small change in the constraint level K (e.g., δK could be set equal to the error in the side constraint at the kth iteration). Now to first order, δK is given by $\delta K = M_\alpha\, \delta\alpha$. Substituting $\delta\alpha$ from Equation (2.53) gives

$$\begin{aligned} \delta K &= \epsilon M_\alpha (L_\alpha + (\nu^k)^{\mathrm{T}} M_\alpha)^{\mathrm{T}} |_{\alpha=\alpha^k} \\ &= \epsilon(M_\alpha L_\alpha{}^{\mathrm{T}} + M_\alpha M_\alpha{}^{\mathrm{T}} \nu^k)|_{\alpha=\alpha^k} \end{aligned} \tag{2.54}$$

and so ν^k is given by

$$\nu^k = (M_\alpha M_\alpha^{\mathrm{T}})^{-1} (\delta K/\epsilon - M_\alpha L_\alpha^{\mathrm{T}}) \tag{2.55}$$

As α^k approaches the optimal solution it may be verified easily that ν^k approaches the Lagrange multiplier ν.†

2.3.2 Summary of the Algorithm

1. Choose a nominal control parameter α^k and evaluate the value of the performance index

$$J^k = L(x, \alpha^k) \tag{2.56}$$

2. For some step size ϵ^k form α^{k+1}

$$\alpha^{k+1} = \alpha^k + \epsilon^k L_\alpha^{\mathrm{T}}(x, \alpha^k) \tag{2.57}$$

and evaluate the return J^{k+1}.

3. If $J^{k+1} > J^k$ set $k = k + 1$ and go to step 2.

4. If $J^{k+1} \leqslant J^k$ set $\epsilon^k = \epsilon^k/\beta$ where β is some arbitrary constant greater than unity. (In practice $\beta = 2$ is satisfactory.)

5. Repeat steps 2, 3, and 4 until the required accuracy is attained.

6. If constraints are present, Equations (2.56) and (2.57) are replaced by

$$J^* = L(x, \alpha^k) + \nu^k M(x, \alpha^k) \tag{2.58}$$

$$\alpha^{k+1} = \alpha^k + \epsilon^k (L_\alpha + (\nu^k)^{\mathrm{T}} M_\alpha)^{\mathrm{T}} \tag{2.59}$$

where

$$\nu^k = (M_\alpha M_\alpha^{\mathrm{T}})^{-1} (\delta K/\epsilon - M_\alpha L_\alpha^{\mathrm{T}}) \tag{2.60}$$

and where δK is determined so that the error in the constraint will be decreased.

2.3.3 Numerical Examples

EXAMPLE 2.4. Consider Example 2.2 in which the lift L of a foil suspended in a gas was maximized. The lift L was given by

$$L = \sin^2 \alpha \cos \alpha \qquad (2.61)^{‡}$$

† This procedure is often referred to as the *gradient projection method.*

‡ The constant K will be taken to be unity.

The gradient algorithm, essentially Equations (2.56) and (2.57), was programmed on the 7090 computer with the gradient L_α given by

$$L_\alpha = 2 \sin \alpha \cos^2 \alpha - \sin^3 \alpha \tag{2.62}$$

The nominal value for α^k was chosen to be $\alpha = 0.0001$, and the resulting sequence of control parameters converged fairly rapidly to the optimum as computed in Example 2.2. This convergence is shown in Table I. The convergence in this case is quite rapid. Unfortunately this

TABLE I. CONVERGENCE DETAILS FOR EXAMPLE 2.4

Iteration Number	α deg	α sec	Lift	Step Length
1	0	1	0.89999991D–07	1.0
2	0	3	0.80999927D–06	1.0
3	0	9	0.72899410D–05	1.0
4	0	27	0.65605217D–04	1.0
5	1	23	0.59010274D–03	1.0
6	4	10	0.52831426D–02	1.0
7	12	26	0.45368885D–01	1.0
8	35	25	0.27379393D 00	1.0
9	68	22	0.31850359D 00	1.0
10	52	35	0.38330343D 00	0.5
11	55	1	0.38486980D 00	0.5
12	54	41	0.38489943D 00	0.5
13	54	44	0.38490016D 00	0.5

is not a characteristic of gradient methods, as will be demonstrated in the next example.

EXAMPLE 2.5. The problem is to minimize the function L where

$$L = 100(\alpha_2 - \alpha_1^2)^2 + (1 - \alpha_1)^2 \tag{2.63†}$$

from an initial guess $\alpha_1^k = 1.2$, $\alpha_2^k = 1.0$. It is interesting to examine Figure 2.3 which shows contours of constant values of L. Obviously the problem is difficult because of the long narrow valley, as the gradient procedure will tend to choose values for α^k alternately on one side and then on the other.

† This problem was originally suggested by Rosenbrock (1960).

The gradient $[L_{\alpha_1}, L_{\alpha_2}]$ is given by

$$
\begin{aligned}
L_{\alpha_1} &= -400\alpha_1(\alpha_2 - \alpha_1^2) - 2(1 - \alpha_1) \\
L_{\alpha_2} &= 200(\alpha_2 - \alpha_1^2)
\end{aligned} \tag{2.64}
$$

The convergence is shown in Table II. It is very slow. In fact the optimal solution $\alpha_1^{\text{opt}} = 1$, $\alpha_2^{\text{opt}} = 1$ is never reached. This slow convergence is a characteristic of the gradient procedure, although this is a somewhat

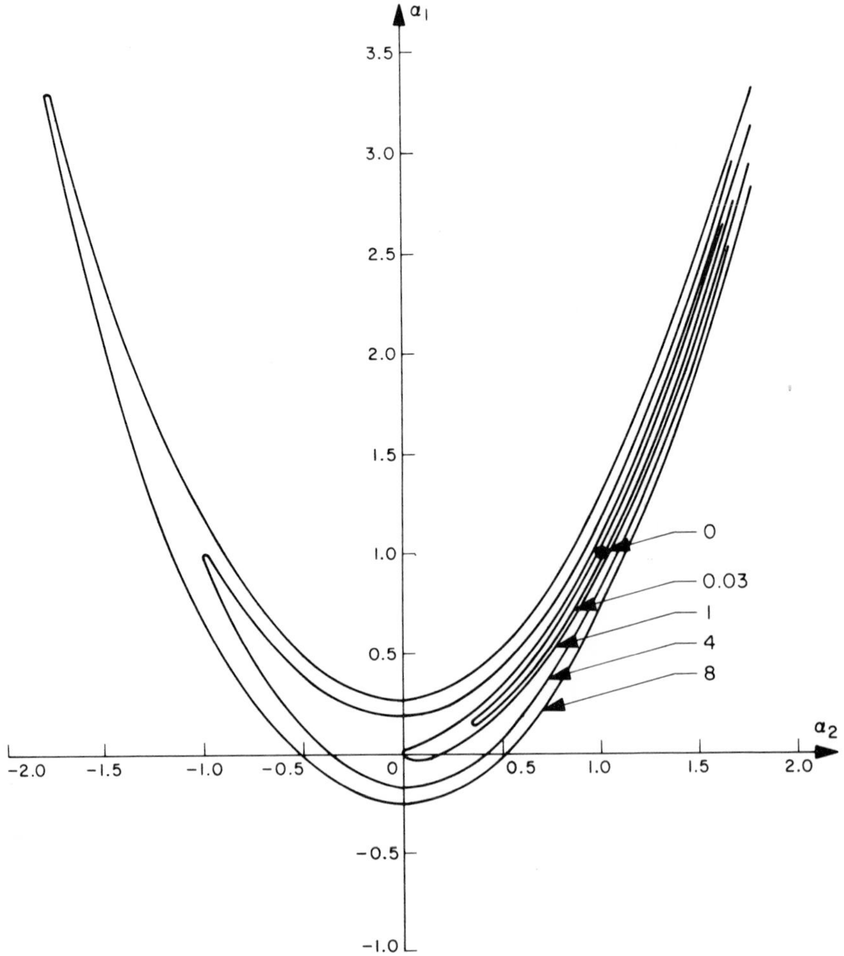

FIG. 2.3. Contours of constant cost for Rosenbrock's function (Example 2.5).

extreme case. Hence the need for an algorithm that gives somewhat faster convergence. One such algorithm is the Newton–Raphson method.

TABLE II. CONVERGENCE DETAILS FOR EXAMPLE 2.5

Iteration Number	α_1	α_2	L	Step Length
1	−0.93681641D−00	0.11074219D−01	0.90319191D−01	0.12207031D−02
2	−0.10974363D−01	0.10176575D−01	0.78852600D−01	0.19531250D−02
5	−0.10099699D−01	0.10479187D−01	0.41177055D−01	0.19531250D−02
10	0.12679947D−01	0.15944683D−01	0.89622972D−01	0.97656250D−03
20	0.12623585D−01	0.15936366D−01	0.68832726D−01	0.19531250D−02
30	0.12611768D−01	0.15906625D−01	0.68214217D−01	0.19531250D−02
40	0.12599884D−01	0.15877038D−01	0.67595736D−01	0.19531250D−02
50	0.12587884D−01	0.15851640D−01	0.67009345D−01	0.97656250D−03
60	0.12577350D−01	0.15823758D−01	0.66450224D−01	0.97656250D−03
70	0.12567006D−01	0.15795913D−01	0.65903891D−01	0.19531250D−02
80	0.12555027D−01	0.15766842D−01	0.65297419D−01	0.19531250D−02
90	0.12543086D−01	0.15737882D−01	0.64697676D−01	0.97656250D−03
100	0.12531218D−01	0.15709019D−01	0.64105174D−01	0.97656250D−03

2.3.4 The Newton–Raphson Algorithm

The gradient method is based upon a first-order expansion of the performance index, whereas the Newton–Raphson method is developed from a second-order expansion of the performance index J^{k+1}, i.e.,

$$J^{k+1} = J^k + L_\alpha(x, \alpha^k)\,\delta\alpha + \tfrac{1}{2}\,\delta\alpha^{\mathrm{T}} L_{\alpha\alpha}(x, \alpha^k)\,\delta\alpha + \cdots \tag{2.65}$$

In a strictly convex region near the optimum [i.e., $L_{\alpha\alpha}(x, \alpha^k) < 0$] this expansion has a unique maximum with respect to $\delta\alpha$ ($\delta\alpha = \alpha^{k+1} - \alpha^k$ must, of course, be chosen so that the expansion is valid), given by

$$\delta\alpha = -L_{\alpha\alpha}^{-1} L_\alpha{}^{\mathrm{T}} \big|_{\alpha=\alpha^k} \tag{2.66}$$

With this value for $\delta\alpha$ it is obvious that J^{k+1} will be greater than J^k.

Thus the Newton–Raphson algorithm generates a sequence of nominal solutions $\alpha^0, \alpha^1, \ldots, \alpha^k$ that converges to the optimal solutions with the algorithm.

$$\alpha^{k+1} = \alpha^k - \epsilon L_{\alpha\alpha}^{-1} L_\alpha{}^{\mathrm{T}} \big|_{\alpha=\alpha^k} \tag{2.67}$$

The parameter ϵ is chosen to ensure the validity of the expansion Equation (2.65). Near to the optimum, it should be set equal to unity.

A geometrical interpretation of the Newton–Raphson algorithm is given in Figure 2.4.

That the performance index is increased by this choice of α^{k+1} may be seen by substituting $\delta\alpha = \alpha^{k+1} - \alpha^k$ into the expansion, Equation (2.65), which, with $\epsilon = 1$ gives

$$J^{k+1} - J^k = -\tfrac{1}{2} L_\alpha L_{\alpha\alpha} L_\alpha^{\mathrm{T}} \tag{2.68}$$

Hence, $J^{k+1} > J^k$ so long as (1) $L_{\alpha\alpha} < 0$ and (2) $\delta\alpha = -L_{\alpha\alpha}^{-1} L_\alpha^{\mathrm{T}}$ is small.

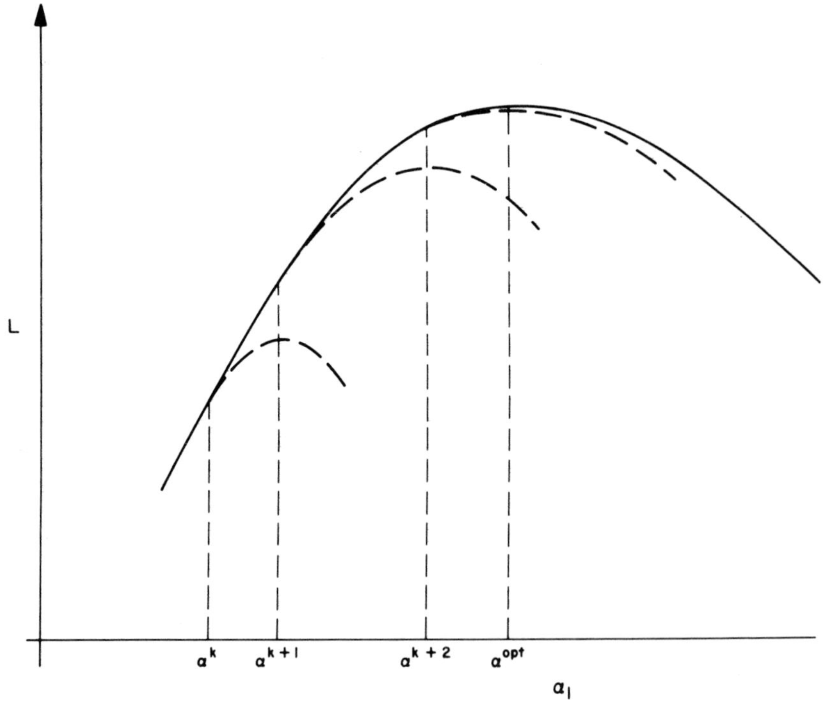

FIG. 2.4. The Newton–Raphson algorithm.

If the second condition does not hold, the step parameter ϵ^k is introduced where $0 < \epsilon < 1$, and Equation (2.81) becomes

$$J^{k+1} - J^k \simeq (\epsilon^2 - 2\epsilon/2) L_\alpha L_{\alpha\alpha}^{-1} L_\alpha^{\mathrm{T}}$$

Now ϵ may always be chosen so that the expansion, Equation (2.65), is a good approximation. But $\epsilon^2 - 2\epsilon$ is always negative for $0 < \epsilon < 1$, and hence the right-hand side is positive. Thus J^{k+1} will be greater than J^k.

However, if condition (1) does not hold, either the gradient method, or, as is often more profitable, a combination of the gradient and Newton–Raphson techniques should be used.

Convergence of the Newton–Raphson method is, as may be expected, much faster than the gradient method. It was shown before that the error in the control parameter after k iterations was e^{k+1}, where, from Equation (2.51), $e^{k+1} = F_\alpha e^k + O(e^k)^2$. In the case of the Newton–Raphson algorithm, F_α is given by $F_\alpha = I(1 - \epsilon)$, and, near the optimum, choosing $\epsilon^k = 1$, e^{k+1} is given by

$$e^{k+1} = O(e^k)^2 \tag{2.69}$$

Thus e^{k+1} is proportional to $(e^k)^2$, and the method is said to have second-order or quadratic convergence. In fact, if the performance index is second order, convergence is obtained in one step.

If side constraints, of the form given by Equation (2.5), are present, the performance index must be modified by adjoining the side constraints with Lagrange multipliers to form an augmented performance index J^*

$$J^* = L(x, \alpha) + \nu^T M(x, \alpha) \tag{2.70}$$

Now both ν and α must be chosen to extremize J^*. Expanding J^* to second order in both α and ν gives

$$J^{*k+1} - J^{*k}$$
$$= (L_\alpha + \nu^T M_\alpha)\,\delta\alpha + M(x, \alpha)\,\delta\nu + \delta\nu^T M_\alpha\,\delta\alpha + \tfrac{1}{2}\,\delta\alpha^T(L_{\alpha\alpha} + \nu^T M_{\alpha\alpha})\,\delta\alpha$$

The Newton–Raphson corrections are obtained by maximizing with respect to α and extremizing with respect to ν. The reader can easily verify that the best choice for $\delta\alpha$ and $\delta\nu$ is

$$\begin{bmatrix} \delta\alpha \\ \delta\nu \end{bmatrix} = -\begin{bmatrix} L_{\alpha\alpha} + \nu^T M_{\alpha\alpha} & M_\alpha \\ M_\alpha^T & 0 \end{bmatrix}^{-1} \begin{bmatrix} L_\alpha + \nu^T M_{\alpha\alpha} \\ M \end{bmatrix}$$

and α^{k+1} and ν^{k+1} are given by

$$\alpha^{k+1} = \alpha^k + \delta\alpha \qquad \text{and} \qquad \nu^{k+1} = \nu^k + \delta\nu$$

Note that the matrix

$$\begin{bmatrix} L_{\alpha\alpha} + \nu^T M_{\alpha\alpha} & M_\alpha \\ M_\alpha^T & 0 \end{bmatrix}$$

will be nonsingular as long as the system is normal (cf Section 2.2.2). The strengthened convexity condition must also hold.

2.3.5 Another Derivation of the Newton–Raphson Algorithm. Direct Methods versus Indirect Methods

In both the derivation of the gradient method and the Newton–Raphson method, the original optimization problem was replaced with a succession of simpler optimization problems. In the gradient method, the simpler problem was obtained by a linearization of the original problem, whereas the Newton–Raphson algorithm was obtained by considering a second-order expansion of the original problem. Such techniques are called direct methods. Another class of methods for solving optimization problems is the class of indirect methods. Indirect methods are obtained from a consideration of the theoretical conditions for an optimum. Examples of indirect methods have already been given in Examples 2.1 and 2.2. Here the theory was used to obtain closed-form solutions to the problem. Numerical techniques, particularly the Newton–Raphson method, may also be developed as an indirect method.

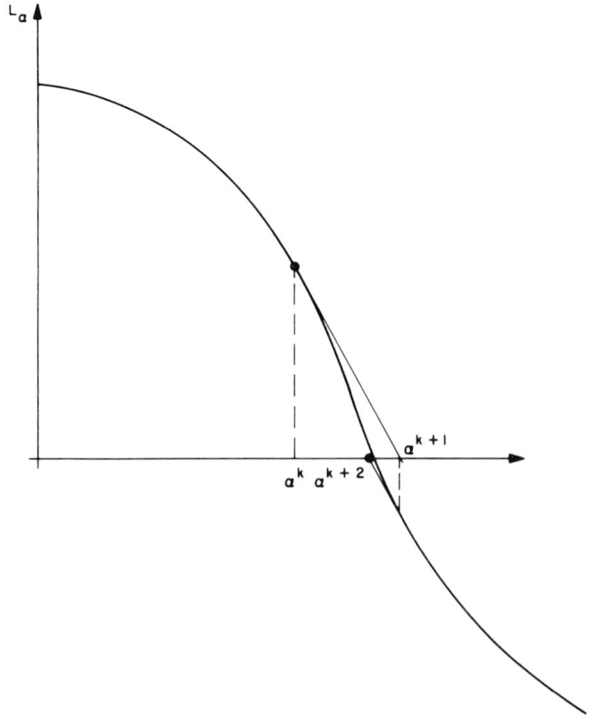

FIG. 2.5. The Newton–Raphson algorithm (indirect approach).

A classical example is the following derivation of the Newton–Raphson method.

The starting point for all indirect derivations of the Newton–Raphson method is the stationarity condition, e.g., for the unconstrained problem $L_\alpha = 0$. Now the original problem of optimizing the index L is replaced by the problem of choosing the control parameters α such that the stationarity condition is satisfied. Suppose that a nominal parameter α^k has been chosen near to the optimal. A variation in the stationarity condition is given by $\delta L_\alpha = L_{\alpha\alpha}\,\delta\alpha$, where

$$\delta L_\alpha = [L_\alpha|_{\alpha=\alpha^{k+1}} - L_\alpha|_{\alpha=\alpha^k}]$$

and $\delta\alpha = \alpha^{k+1} - \alpha^k$. Inverting this relation we obtain

$$\delta\alpha = L_{\alpha\alpha}^{-1}\,\delta L_\alpha$$

But, if the nominal is close to the optimum, $L_\alpha|_{\alpha=\alpha^{k+1}}$ is very small. Thus it is reasonable to choose $\delta\alpha$ by

$$\delta\alpha = -L_{\alpha\alpha}^{-1} L_\alpha|_{\alpha=\alpha^k}$$

This algorithm is, of course, identical to the one derived in Section 2.3.4.

The algorithm, which employs a linear approximation to L_α, is illustrated in Figure 2.5 as a one-dimensional example.

2.3.6 Summary of the Newton–Raphson Algorithm

1. Choose a nominal control parameter α^k and evaluate the performance index

$$J^k = L(x, \alpha^k) \tag{2.71}$$

2. Set the step size $\epsilon^k = 1$ and form α^{k+1}

$$\alpha^{k+1} = \alpha^k - \epsilon^k L_{\alpha\alpha}^{-1} L_\alpha \tag{2.72}$$

Note: Check that $L_{\alpha\alpha}$ is negative definite.

3. If $J^{k+1} > J^k$, set $k = k + 1$ and go to step 2.
4. If $J^{k+1} \leqslant J^k$, set $\epsilon^k = \epsilon^k/2$ and return to step 2.
5. Repeat steps 2, 3, and 4 until the desired accuracy is attained.
6. If constraints are present, Equations (2.71) and (2.72) must be replaced by

$$J^{*k} = L(x, \alpha^k) + \nu^{kT} M(x, \alpha^k) \tag{2.73}$$

and

$$\begin{bmatrix} \alpha^{k+1} \\ \nu^{k+1} \end{bmatrix} = \begin{bmatrix} \alpha^{k} \\ \nu^{k} \end{bmatrix} - \epsilon^k \begin{bmatrix} L_{\alpha\alpha} + \nu^{\mathrm{T}} M_{\alpha\alpha} & M_\alpha^{\mathrm{T}} \\ M_\alpha & 0 \end{bmatrix}^{-1} \begin{bmatrix} L_\alpha^{\mathrm{T}} + M_\alpha^{\mathrm{T}} \nu^k \\ M \end{bmatrix} \tag{2.74}$$

2.3.7 Examples

The two examples used to illustrate the gradient method are repeated so that a comparison of the convergence properties may be made.

EXAMPLE 2.6. Maximization of the lift L where $L = \sin^2 \alpha \cos \alpha$ and the gradient L_α is $L_\alpha = 2 \sin \alpha \cos^2 \alpha - \sin^3 \alpha$. The second derivative $L_{\alpha\alpha}$ is $L_{\alpha\alpha} = 2 \cos^3 \alpha - 7 \sin^2 \alpha \cos \alpha$.

For the same initial guess for α^k, the convergence is given in Table III.

TABLE III. CONVERGENCE DETAILS FOR EXAMPLE 2.6

Iteration Number	α deg	α sec	Lift	Step Length ϵ
1	0	1	0.89999991D–07	1.0
2	0	3	0.80999927D–06	1.0
3	0	9	0.72899410D–05	1.0
4	0	27	0.65605217D–04	1.0
5	1	23	0.59010274D–03	1.0
6	4	10	0.52831426D–02	1.0
7	12	26	0.45368885D–01	1.0
8	35	25	0.27379393D–00	1.0
9	55	10	0.38483425D–00	0.5
10	54	44	0.38490018D–00	1.0

As can be seen, the solution is found after ten iterations, which is somewhat better than with the gradient method. However, a more substantial improvement is obtained with the second example.

EXAMPLE 2.7. The second problem was to minimize the function L, $L = 100(\alpha_2 - \alpha_1^2)^2 + (1 - \alpha_1)^2$, with an initial guess of $\alpha_1 = -1.2$ and $\alpha_2 = 1.0$. The first partial derivatives are

$$L_{\alpha_1} = -400\alpha_1(\alpha_2 - \alpha_1^2) - 2(1 - \alpha_1)$$

$$L_{\alpha_2} = 200(\alpha_2 - \alpha_1^2)$$

and the second partial derivatives are

$$L_{\alpha_1\alpha_1} = -400(\alpha_2 - \alpha_1^2) + 800\alpha_1^2 + 2$$
$$L_{\alpha_2\alpha_2} = 200$$
$$L_{\alpha_1\alpha_2} = -400\alpha_1$$

The convergence is shown in Table IV. As may be seen, the convergence is rapid and the exact (to 16 significant figures) solution has been obtained. This is in marked contrast to the solution obtained with the gradient method.

TABLE IV. CONVERGENCE DETAILS FOR EXAMPLE 2.7

Iteration Number	α_1	α_2	L	Step Length
1	−0.11752809D–01	0.13806742D–01	0.47318843D–01	0.100D–01
2	−0.93298143D–00	0.81121066D–00	0.40873987D–01	0.125D–00
3	−0.78254008D–00	0.58973638D–00	0.32286726D–01	0.100D–01
4	−0.45999712D–00	0.10756339D–00	0.32138981D–01	0.100D–01
5	−0.39304563D–00	0.15000237D–00	0.19425854D–01	0.100D–01
6	−0.20941191D–00	0.67701267D–02	0.16001937D–01	0.250D–00
7	−0.65719021D–01	−0.16328656D–01	0.11783896D–01	0.100D–01
8	0.14204255D–00	−0.22988784D–01	0.92241158D–00	0.100D–01
9	0.23110720D–00	0.45478024D–01	0.59748862D–00	0.100D–01
10	0.37974282D–00	0.11814580D–00	0.45262510D–00	0.500D–00
11	0.47959489D–00	0.22004082D–00	0.28076244D–00	0.100D–00
12	0.65340583D–00	0.39672894D–00	0.21139340D–00	0.100D–01
13	0.70262363D–00	0.49125758D–00	0.89019501D–01	0.100D–01
14	0.80278553D–00	0.63322101D–00	0.51535405D–01	0.500D–00
15	0.86349081D–00	0.74193125D–00	0.19992778D–01	0.100D–01
16	0.94207869D–00	0.88133620D–00	0.71692436D–02	0.100D–01
17	0.96799182D–00	0.93633667D–00	0.10696137D–02	0.100D–01
18	0.99621031D–00	0.99163870D–00	0.77768464D–04	0.100D–01
19	0.99947938D–00	0.99894834D–00	0.28246695D–06	0.100D–01
20	0.99999889D–00	0.99999751D–00	0.85170750D–11	0.100D–01
21	0.10000000D–01	0.10000000D–01	0.37439893D–20	0.100D–01
22	0.10000000D–01	0.10000000D–01	0.32047474D–30	0.100D–01
23	0.10000000D–01	0.10000000D–01	0.00000000D–38	0.100D–01

2.3.8 Conjugate Direction Methods

There are many computational schemes other than the gradient or Newton–Raphson algorithms that have been devised for parameter optimization. It is beyond the scope of this book to discuss all of these methods, and the interested reader should consult the references listed in the bibliography at the end of this chapter.

However, one particular class of methods, the so-called conjugate direction or conjugate gradient methods, which were originally developed for minimizing quadratic forms, have recently been shown to be particularly useful and will be discussed briefly. These methods are claimed to exhibit a speed of convergence that approaches that of the Newton–Raphson algorithm without the need for the formation or inversion of the matrix of second partial derivatives $L_{\alpha\alpha}$. There are several different versions of these algorithms, although only two will be given here: the basic conjugate gradient algorithm [Hestenes and Stiefel (1952)] and the Davidon algorithm [Davidon (1959)]. First, however, the conjugate direction process will be defined.

Consider a quadratic function $L(\alpha)$ that is assumed to have a matrix of second partial derivatives which is negative definite. A set of distinct nonzero vectors $\beta^1, \beta^2, \ldots, \beta^p$ is said to be conjugate with respect to $L_{\alpha\alpha}$ if they have the property that

$$\beta^{i\mathrm{T}} L_{\alpha\alpha} \beta^j = 0, \qquad i \neq j \tag{2.75}$$

Thus the vectors β^i are linearly independent and form a basis in the p-dimensional Euclidian space. Hence, any arbitrary p-dimensional vector α may be defined in terms of the conjugate set β^i

$$\alpha = \sum_{i=1}^{p} c_i \beta^i \tag{2.76}$$

where the c_i are unknown *scalar* parameters that must be determined.

For example, suppose $L(\alpha)$ has the form

$$L(\alpha) = Q\alpha + \tfrac{1}{2}\alpha^{\mathrm{T}} R \alpha \tag{2.77}$$

where Q is a p-dimensional row vector and R is a $p \times p$ negative definite matrix. Substituting the foregoing expression, (2.76), for α in Equation (2.77), L becomes

$$L(\alpha) = Q \sum_{i=1}^{p} c_i \beta^i + \tfrac{1}{2} \left(\sum_{i=1}^{p} c_i \beta^i \right)^{\mathrm{T}} R \left(\sum_{i=1}^{p} c_i \beta^i \right)$$

which, because of the conjugacy of the β^i, may be written

$$L = Q \sum_{i=1}^{p} c_i \beta^i + \tfrac{1}{2} \sum_{i=1}^{p} [c_i^2 \beta^{i\mathrm{T}} R \beta^i]$$

$$= \sum_{i=1}^{p} \{\tfrac{1}{2}\lambda(i)\, c_i^2 + r(i)\, c_i\}$$

where the scalars $\lambda(i)$ and $r(i)$ are given by

$$r(i) = Q\beta^i, \qquad i = 1, 2, \ldots, \rho$$
$$\lambda(i) = \beta^{iT}R\beta^i,$$

Now the c_i may be found by maximizing L with respect to the c_i

$$c_i = -r(i)/\lambda(i), \qquad i = 1, 2, \ldots, \rho$$

The main advantage of employing a conjugate basis is, of course, that the parameters $c_1, c_2, \ldots, c_\rho$ may be found independently, thus avoiding the inversion of the matrix $L_{\alpha\alpha} = R$. For nonlinear problems, in order to avoid evaluating $L_{\alpha\alpha}$, a succession of one-dimensional search procedures may be used to find the maximizing values of the c_i. A typical procedure would be as follows.

1. Choose α^k by

$$\alpha^k = \alpha^{k-1} + c_k\beta^k = \alpha^0 + \sum_{i=1}^{k} c_i\beta^i$$

where α^0 is the nominal solution.

2. Choose c_{k+1} such that $J = L(\alpha^k + c_{k+1}\beta^{k+1})$ is maximized.
3. Set $k = k + 1$ and go to 1.

For a linear quadratic problem convergence will be obtained in ρ steps for a ρ-dimensional problem.

The different conjugate direction algorithms are distinguished by the manner in which the conjugate directions β^i are chosen. The classic algorithm, given by Hestenes and Stiefel (1952) and Beckman (1960), is as follows.

1. α^k is chosen by

$$\alpha^k = \alpha^{k-1} + c_k\beta^k, \qquad \beta^1 = L_\alpha{}^T|_{\alpha=\alpha^k}$$

where β^k is given by

$$\beta^k = L_\alpha{}^T(\alpha^k) + \frac{\| L_\alpha(\alpha^k)\|^2}{\| L_\alpha(\alpha^{k-1})\|^2}\beta^{k+1}, \qquad k \geqslant 2$$

and where α^0 is the priming solution.

2. c_k is chosen to maximize $L(\alpha^k)$.

Note that although convergence should be obtained in ρ steps for the quadratic problem, convergence for the general problem may be much slower.

Another conjugate direction algorithm is that given by Davidon (1959) and Fletcher and Powell (1963),

$$\alpha^k = \alpha^{k-1} + \Delta\alpha^k$$

where

$$\Delta\alpha^k = c_k H^k L_\alpha{}^{\mathrm{T}} |_{\alpha=\alpha^{k-1}}$$

The weighting matrix H is updated by the equations

$$\begin{aligned} H^{k+1} &= H^k + A^k + B^k \qquad [H' = I] \\ A^k &= (\Delta\alpha^k)(\Delta\alpha^k)^{\mathrm{T}} \\ B^k &= -H^k(\Delta L_\alpha{}^k)^{\mathrm{T}} (\Delta L_\alpha{}^k)\, H^k / \Delta L_\alpha{}^k H^k (\Delta L_\alpha{}^k)^{\mathrm{T}} \\ \Delta L_\alpha{}^k &= L_\alpha |_{\alpha=\alpha^k} - L_\alpha |_{\alpha=\alpha^{k-1}} \end{aligned}$$

where H^1 is specified as a negative definite matrix and c_k is again chosen to maximize $L(\alpha^k)$. (**Note:** Here $\beta^k = H^k L_\alpha|_{\alpha=\alpha^k}$.)

It can be shown that as $k \to p$, the matrix H becomes $-L_{\alpha\alpha}^{-1}$, and hence the correction approaches that of the Newton–Raphson algorithm. Consequently, it might be expected that Davidon's method is somewhat superior in terms of convergence to the conjugate gradient algorithm for nonlinear problems. An interesting property of these methods which is shown by Myers (1968) is that for the second-order problem both techniques generate the same conjugate directions.

Numerical results will not be presented here, and readers are referred to Kelley and Myers (1967) for a comparison of the various conjugate direction methods.

PROBLEMS

1. Consider $J = -x^2 + 4xy - 2y^2 + y$. Find the values of x and y such that J is stationary. At this point, regarding x as fixed, is J a maximum with respect to y? Holding y fixed, is J minimized with respect to x? Prove both assertions.
2. If both x and y are free in Problem 1, is the above point a maximum for J?
3. To Problem 1, add the constraint $2x + 4y = 3$. What is the extremal value? Is it a maximum or a minimum? Prove it.
4. A pole of length a is suspended vertically such that its lower end is a height k above the floor. Find the point on the floor from which the pole subtends the greatest angle.
5. Repeat Example 2.1 for the case in which x is an n-dimensional vector and α is a p-dimensional vector.

The following problems are given by Fletcher and Powell (1963) for numerical computation.

6. Minimize the function L where

$$L = (\alpha_1 + 10\alpha_2)^2 + 5(\alpha_3 - \alpha_4)^2 + (\alpha_2 - 2\alpha_3)^4 + 10(\alpha_1 - \alpha_4)^4$$

with a nominal $\alpha = [3, -1, 0, 1]$ using the numerical techniques described in this chapter.

7. Minimize the function L where

$$L = 100\{[\alpha_3 - 10\theta]^2 + [r - 1]^2 + \alpha_3{}^2\}$$

where

$$\begin{aligned} \theta &= \frac{1}{2\pi}\arctan\left(\frac{\alpha_2}{\alpha_1}\right), & \alpha_1 > 0 \\ &= \frac{1}{2} + \frac{1}{2\pi}\arctan\left(\frac{\alpha_2}{\alpha_1}\right), & \alpha_1 < 0 \end{aligned}$$

and

$$r = (\alpha_1{}^2 + \alpha_2{}^2)^{1/2}$$

Values of θ and α_3 should only be considered for $-\pi/2 < 2\pi\theta < 3\pi/2$ and $-2.5 < \alpha_3 < 7.5$. The nominal value of α is $[-1, 0, 0]$. This function has a helical valley in the α_3 direction with pitch 10 and radius 1.

8. Using an indirect method, derive the Newton–Raphson algorithm for problems with side constraints.

BIBLIOGRAPHY AND COMMENTS

Section 2.2. For a more detailed discussion of the theory of maxima and minima, the reader is referred to the following texts. An exhaustive list of classical references may be found in Hancock (1960).

Courant, R. and Hilbert, D. (1961). *Methods of Mathematical Physics.* Wiley (Interscience), New York.

Caratheodory, C. (1967). *Calculus of Variations and Partial Differential Equations of the First Order* (J. J. Brandstatter, ed. and R. Dean, transl.), Vol. 2, Chap. 11. Holden-Day, San Francisco, California.

Edelbaum, T. N. (1962). "Theory of Maxima and Minima," *Optimization Techniques* (G. Leitmann, ed.), pp. 1–32. Academic Press, New York.

Hancock, H. (1960). *Theory of Maxima and Minima (1917).* Dover, New York.

Hestenes, M. R. (1966). *Calculus of Variations and Optimal Control Theory*, Chap. 1. Wiley, New York.

Section 2.3. The material in this section is primarily intended to provide a background for the following chapters. Readers interested in parameter optimization may find the following books provide a more detailed discussion.

Booth, A. D. (1957). *Numerical Methods*. Butterworth, London and Washington, D.C.
Scarborough, J. B. (1966). *Numerical Mathematical Analysis*. Johns Hopkins Press, Baltimore, Maryland.

Example 2.5. This example was suggested by Rosenbrock (1960) to illustrate an optimization technique that avoids the explicit evaluation of the gradient L_α.

Rosenbrock, H. H. (1960). "An Automatic Method for Finding the Greatest or Least Value of a Function," *Computer J.*, Vol. 3, p. 175.

Section 2.3.8. Over the last few years several papers have been published that deal exclusively with conjugate direction methods. In particular, the paper by Kelley and Myers (1967) effects a comparison of several versions of the procedure.

Beckman, F. S. (1960). "The Solution of Linear Equations by the Conjugate Gradient Method," *Mathematical Methods for Digital Computers* (A. Ralston and H. S. Wilf, eds.). Wiley, New York.
Booth, A. D. (1957). *Numerical Methods*. Butterworth, London and Washington, D.C.
Davidon, W. C. (1959). "Variable Metric Method for Minimization," *Argonne National Laboratory Report* ANL-5990, Rev.
Fletcher, R. and Powell, M. J. D. (1963). "A Rapidly Convergent Descent Method for Minimization," *Computer J.*, Vol. 6, p. 163.
Fletcher, R. and Reeves, C. M. (1964). "Function Minimization by Conjugate Gradients," *Computer J.*, Vol. 6, p. 149.
Hestenes, M. R. and Stiefel, E. (1952). "Methods of Conjugate Gradients for Solving Linear Systems," *J. Res. Natl. Bur. Std.*, Vol. 49, p. 409.
Hestenes, M. R. (1956). "The Conjugate-Gradient Method for Solving Linear Systems," *Proc. Symp. Appl. Math.*, Vol. 6. McGraw-Hill, New York.
Kelley, H. J., Denham, W. F., Johnson, I. L., and Wheatley, P. O. (1966). "An Accelerated Gradient Method for Parameter Optimization with Nonlinear Constraints," *J. Astronaut. Sci.*, Vol. XIII, No. 4, pp. 166–169.
Kelley, H. J. and Myers, G. E. (1967). "Conjugate Direction Methods for Parameter Optimization." Presented at the 18th Congress of the International Astronautical Federation, Belgrade, Yugoslavia.
Myers, G. E. (1968). "Properties of the Conjugate Gradient and Davidon Methods," *J. Optimization Theory Appl.*, Vol. 2, No. 4, p. 209.

Wilde and Beightler (1967) provide a wealth of material and many references to the techniques of linear and nonlinear programming which have not been discussed here. These techniques are particularly useful for optimization problems characterized by an abundance of inequality constraints.

Wilde, D. J. and Beightler, C. S. (1967). *Foundations of Optimization*. Prentice-Hall, Englewood Cliffs, New Jersey.

3

Optimal Control of Discrete Systems

The problems discussed in the last chapter were characterized by the choice of certain parameters in order to optimize some performance index. In this chapter the problems considered will involve such choices or decisions at each of a finite set of times or stages. A cost, or the value of a performance index, will be associated with each sequence of decisions and the problem will be to optimize this cost. This is the so-called multistage, or discrete, optimal control problem.

After defining some notation, the solution of the problem is reformulated in terms of dynamic programming. The solution is shown to depend upon the maximization of a sequence of functions. Next, an analytic solution is shown to exist for the *linear quadratic* problem. This analysis is illustrated with a simple example.

Unfortunately, in general analytic solutions do not exist, and recourse must be made to numerical methods. These methods are the subject of Sections 3.5–3.8. Two methods, the gradient and the Newton–Raphson (successive sweep), are described in detail and are shown to lead to stationarity and convexity conditions for a local maximum of the multistage problem. The neighboring extremal method and its relation to the successive sweep method is discussed. Finally, necessary and sufficient conditions for a local maximum of the general problem are given.

3.1 NOTATION

3.1.1 System Variables

At each stage the system is assumed to be characterized by a finite set of real numbers that is referred to as the state of the system. The state, at the ith stage, is denoted by an n-dimensional column vector $x(i)$ where

$$x(i) = \begin{bmatrix} x_1(i) \\ x_2(i) \\ \vdots \\ x_n(i) \end{bmatrix} \tag{3.1}$$

The control variables are classified in two sets. Variables of the one set are referred to as *control functions*, or, more simply, as *controls*. These control functions, which normally vary from stage to stage, are denoted by an m-dimensional column vector of real numbers $u(i)$ where

$$u(i) = \begin{bmatrix} u_1(i) \\ u_2(i) \\ \vdots \\ u_m(i) \end{bmatrix} \tag{3.2}$$

The other set of control variables are *constant* and are referred to as *control parameters*. The control parameters are denoted by a p-dimensional vector α

$$\alpha = \begin{bmatrix} \alpha_1 \\ \alpha_2 \\ \vdots \\ \alpha_p \end{bmatrix} \tag{3.3}$$

3.1.2 System Dynamics

The dynamics of the system are expressed in terms of a set of discrete equations. The process is assumed to have N stages, and the evolution of the state through these stages, $x(0), x(1), \ldots, x(N)$ is governed by an equation of the form

$$x(i+1) = F(x(i), u(i), \alpha), \qquad i = 0, 1, \ldots, N-1 \tag{3.4}$$

where $F = (F_1, F_2, \ldots, F_n)^{\mathrm{T}}$† is an n-dimensional vector of functions.

3.1.3 Performance Index

The performance index is denoted by J,

$$J = \phi(x(N), \alpha) + \sum_{i=0}^{N-1} L(x(i), u(i), \alpha) \tag{3.5}$$

† Superscript T is used to denote the transpose.

where L and ϕ are scalar, single-valued functions of their respective arguments. The total number of stages N may be free or specified. In some cases, especially in stabilization problems, N may be regarded as being infinite.

3.1.4 Side Constraints

The most general form of constraint is given by

$$0 = \psi(x(N), \alpha) + \sum_{i=0}^{N-1} M(x(i), u(i), \alpha) \tag{3.6}$$

where ψ and M may be scalar-valued or vector functions. If the function $M = 0$, the foregoing is referred to as a *terminal constraint*. As well as this *equality* constraint, inequality constraints may also have to be considered. In particular, constraints on the state of the form

$$S(x(i)) \leqslant 0 \tag{3.7}$$

and on the control variables,

$$C(x(i), u(i)) \leqslant 0 \tag{3.8}$$

are very common. The functions C and S may be scalar-valued or vector functions. Constraints may also be placed on the initial state, although in the analysis that follows the initial state is assumed specified.

3.2 THE PROBLEM

In the light of these definitions it is now possible to define the problem in a precise manner, viz: To find the sequence of controls $u(i)$, $i = 0, 1, \ldots, N-1$, and the control parameters α that maximize the performance index, Equation (3.5), subject to the state transition equation (3.4) and the constraints, for example, Equations (3.6)–(3.8).

3.3 DYNAMIC PROGRAMMING SOLUTION

The dynamic programming solution is based on principles that are a direct consequence of the structure of the problem.

3.3.1 The Principle of Causality

This principle is a fundamental property of deterministic multistage systems. It may be stated as follows. *Principle of Causality*: The state $x(k)$ and control parameter α at the kth stage, together with the sequence of controls $u[k, r-1] = [u(k), u(k+1), \ldots, u(r-1)]$ uniquely determine the state of the rth stage $x(r)$. The principle follows directly from a consideration of the system equations (3.4).

The principle implies that there exists a function

$$G(x(k), \alpha, u[k, r-1], k, \alpha)$$

such that

$$x(r) = G(x(k), \alpha, u[k, r-1], k, r) \tag{3.9}$$

G is the *transition function.* Thus the initial state $x(0)$, control parameters α, and the control sequence $u[0, N-1]$ uniquely determine the trajectory $x[1, N] = [x(1), x(2), \ldots, x(N)]$. Hence, the performance index J defined by Equation (3.5) may be written as a function of $x(0)$, α, and $u[0, N-1]$, i.e., there exists a function $V(x(0), \alpha, u[0, N-1])$ such that

$$J = V(x(0), \alpha, u[0, N-1]) \tag{3.10}$$

If $x(0)$ is specified, this relation implies that it is necessary only to determine α and $u[0, N-1]$ to maximize J. This assumption is implicit in the statement of the problem.

3.3.2 The Principle of Optimality

Suppose that somehow the control sequence $u[0, k-1]$ and the control parameter α have been chosen in an optimal manner. Then, from the principle of causality, the trajectory $x[0, k]$ is also determined. Now the performance index J may be written as $J = J_1 + J_2$ where

$$J_1 = \sum_{i=0}^{k-1} L(x(i), \alpha, u(i)) \tag{3.11}$$

and,

$$J_2 = \sum_{i=k}^{N-1} \{L(x(i), \alpha, u(i))\} + \phi(x(N), \alpha) \tag{3.12}$$

The first term, J_1, has been determined from the assumption that the optimal $u[0, k-1]$ and α are known. Thus, to complete the optimization of J, it is clear that it is both necessary and sufficient to determine $u[k, N-1]$ to maximize J_2. This observation was summarized precisely by Bellman as the principle of optimality. It may be stated as follows:

> An optimal sequence of controls in a multistage optimization problem has the property that whatever the initial stage, state, and controls are, the remaining controls must constitute an optimal sequence of decisions for the remaining problem with stage and state resulting from the previous controls considered as initial conditions.

3.3.3 The Principle of Optimal Feedback Control

An important consequence of the principle of optimality is that the choice of the optimal control at some stage may be expressed as a function of the state at that stage. This idea is defined in the principle of optimal feedback control which asserts that "The optimal control at the kth stage, $u(k)$, provided it exists and is unique, may be expressed as a function of the state at the kth stage, $x(k)$, and the control parameter, α.

The principle implies that there exists a function $u^{opt}(x(k), \alpha, k)$ such that

$$u^{opt}(k) = u^{opt}(x(k), \alpha, k) \tag{3.13}$$

This function, $u^{opt}(x(k), \alpha, k)$ is referred to as the *optimal control law*, and, in conjunction with the system equations, may be used to generate optimal control sequences. The optimal control law defines the closed-loop solution to the optimal control problem, and the actual numerical sequence of optimal controls, $u^{opt}(k)$ $(k = 0, 1, \ldots, N-1)$, is referred to as the open-loop solution. The closed-loop solution is more desirable, as it yields the open-loop solutions for *all* values of the initial state and initial stage.

The principle of optimal feedback control may be derived quite easily from a consideration of the optimality and causality principles. First, from the causality principle it follows that there exists a function $V(x(k), \alpha, u[k, N-1], k)$ such that

$$V(x(k), \alpha, u[k, N-1], k) = \phi(x(N), \alpha) + \sum_{i=k}^{N-1} \{L(x(i), u(i), \alpha)\} \tag{3.14}$$

Now, from the principle of optimality $u[k, N-1]$ must be chosen to maximize $V(x(k), \alpha, u[k, N-1], k)$.

Clearly, $u^{opt}[k, N-1]$ [and hence $u^{opt}(k)$] will be functionally dependent upon $x(k)$, α, and k.

3.3.4 The Return Function

The function $V(x(k), \alpha, u[k, N-1], k)$ introduced before is referred to as the *return function* corresponding to the control sequence $u[k, N-1]$. If the control sequence $u[k, N-1]$ is replaced by the optimal control $u^{opt}[k, N-1]$, V becomes the *optimal return function* $V^{opt}(x(k), \alpha, k)$ where

$$V^{opt}(x(k), \alpha, k) = V(x(k), \alpha, u^{opt}[k, N-1], k)$$

The dynamic programming solution will require the construction of the *optimal return function.* Hence it is appropriate here to explain how the return function may be obtained. Let $u(x(i), \alpha, i)$, $i = k + 1, \ldots, N - 1$, be some arbitrary control law and $V(x(k), \alpha, k)$ the corresponding return function where

$$V(x(k), \alpha, k) = \phi(x(N), \alpha) + \sum_{i=k}^{N-1} \{L(x(i), u(x(i), \alpha, i), \alpha, i)\} \tag{3.15}$$

and where $x(k, N)$ is chosen such that

$$x(i + 1) = F(x(i), \alpha, u(x(i), \alpha, i)), \qquad i = k,\ k + 1, \ldots, N - 1 \tag{3.15a}$$

Now, from the definition of the return function, Equation (3.15), $V(x(k), \alpha, k)$ must satisfy the following backward transition equation

$$V(x(k), \alpha, k) = L(x(k), u(x(k), \alpha, k), \alpha) + V(x(k + 1), \alpha, k + 1) \tag{3.16}$$

where $x(k + 1)$ is given by Equation (3.15a) with $i = k$. At the final stage,

$$V(x(N), \alpha, N) = \phi(x(N), \alpha) \tag{3.17}$$

Thus the return function $V(x(i), \alpha, i)$, may be constructed as a backwards sequence on the index i using Equations (3.16) and (3.17). If the general feedback law is replaced by the *optimal control law,* the foregoing equations may be used to construct the *optimal return function,* i.e.,

$$V^{\text{opt}}(x(N), \alpha) = \phi(x(N), \alpha)$$
$$V^{\text{opt}}(x(k), \alpha, k) = L(x(k), \alpha, u^{\text{opt}}(x(k), k)) + V^{\text{opt}}(x(k + 1, \alpha, k + 1) \tag{3.18}$$

where

$$x(k + 1) = F(x(k), \alpha, u^{\text{opt}}(x(k), \alpha, k)) \tag{3.19}$$

3.3.5 Derivation of the Optimal Control Law

Next, the relations that define the optimal control law will be obtained. Let the initial stage be denoted by k, and it is assumed that the optimal control law for the succeeding stages, $u^{\text{opt}}(x(i), \alpha, i)$ $i = k + 1, k + 2, \ldots, N - 1$, has somehow been found. From Equations (3.18) and (3.19) it is clear that this will determine $V^{\text{opt}}(x(k + 1), \alpha, k + 1)$. Now define a return function, $R(x(k), u(k), \alpha, k)$, as the return obtained from an initial stage k and state $x(k)$ by using $u(k)$ followed by $u^{\text{opt}}[k + 1, N - 1]$, i.e.,

$$R(x(k), u(k), \alpha) = L(x(k), u(k), \alpha) + V^{\text{opt}}(x(k + 1), \alpha, k) \tag{3.20}$$

where $x(k+1)$ is given by

$$x(k+1) = F(x(k), u(k), \alpha) \tag{3.21}$$

It follows from the principle of optimality that the optimal choice of $u(k)$ must be such that R is maximized, i.e.,

$$u^{\text{opt}}(x(k), \alpha) = \arg \max_{u(k)} \{R(x(k), u(k), \alpha)\} \tag{3.22}$$

The control parameters α may be solved for at the initial stage by maximizing the optimal return $V^{\text{opt}}(x(0), \alpha)$ with respect to α. In general, the closed-loop solution for α^{opt} can be found as a function of the initial state, i.e.,

$$\alpha^{\text{opt}}(x(0)) = \arg \max_{\alpha} \{V^{\text{opt}}(x(0), \alpha)\} \tag{3.23}$$

These equations [(3.17), (3.22), and (3.23)] completely define the optimal solution. This solution may be summarized as follows.

1. Set

$$V^{\text{opt}}(x(N), \alpha) = \phi(x(N), \alpha) \tag{3.24}$$

2. Given $V^{\text{opt}}(x(k+1), \alpha)$, choose $u^{\text{opt}}(x(k), \alpha)$ from

$$u^{\text{opt}}(x(k), \alpha) = \arg \max_{u(k)} \{R(x(k), u(k), \alpha)\} \tag{3.25}$$

3. Given $u^{\text{opt}}(x(k), \alpha)$ and $V^{\text{opt}}(x(k+1), \alpha)$, form $V^{\text{opt}}(x(k), \alpha)$ from

$$V^{\text{opt}}(x(k), \alpha) = L(x(k), u^{\text{opt}}(x(k), \alpha)) + V^{\text{opt}}(F(x(k), u^{\text{opt}}(x(k), \alpha), \alpha) \tag{3.26}$$

4. Set $k = k - 1$ and repeat steps 2 and 3 until $k = 0$.

5. At $k = 0$ solve for the optimal choice of α by maximizing $V^{\text{opt}}(x(k), \alpha)$.

In utilizing these equations there may be some question as to whether or not a solution exists or as to whether the solution is unique. However, if a unique solution to the dynamic programming equations does exist, then from the preceeding analysis it is clear that a unique optimal solution does exist and is given by the dynamic programming solution.

If side constraints are present, only admissible values of the control variables may be considered in the maximization, Equation (3.25). If the constraint is of the form $C(x(i), u(i)) \leqslant 0$, it is usually satisfied quite easily. A far more difficult type of constraint is

$$\psi(x(N), \alpha) + \sum_{i=0}^{N-1} M(x(i), u(i), \alpha) = 0 \tag{3.27}$$

In general, the best approach, as in the parameter optimization problem, is to adjoin these constraints to the performance index with a Lagrange multiplier ν. Thus, an augmented performance index J^* is formed, viz,

$$J^* = \phi(x(N), \alpha) + \nu^{\mathrm{T}}\psi(x(N), \alpha) + \sum_{i=0}^{N-1} \{L(x(i), u(i), \alpha) + \nu^{\mathrm{T}}M(x(i), u(i), \alpha)\} \tag{3.28}$$

The problem now is to find a control $u[0, N-1]$ and parameter ν that *extremizes* J^*. The control does not necessarily maximize J^*, as shall be seen in later sections.

In general, an analytic solution of the dynamic programming equations is *not* possible, and recourse must be made to numerical techniques. One approach to this numerical problem is given in detail in future sections. Fortunately, one major class of problems may be solved analytically, and it will be discussed in detail in the next section.

3.4 LINEAR QUADRATIC PROBLEMS

A *linear quadratic* problem is characterized by a linear system equation and a quadratic performance index. The linear quadratic problem is important for a variety of reasons. In particular, a simple analytic solution for the closed-loop optimal control for a linear quadratic problem may be found. This control consists of a linear combination of the components of the state vector. Because of this simple form, the problem of the mechanization of optimal control laws for linear systems is relatively straightforward. Of course, practical problems are generally nonlinear, but these problems can often be approximated, realistically, by a linear quadratic form. After deriving the solution, the application will be illustrated with examples.

3.4.1 Linear Quadratic Algorithm

So that the analysis may be simplified, side constraints and control parameters will not be included. Linear constraints will be considered in a later section.

The linear system equations are given by

$$x(i+1) = F_x(i)\,x(i) + F_u(i)\,u(i), \qquad i = 0, 1, \ldots, N-1 \tag{3.29}$$

where, as before, x is an n-dimensional state vector and u is an m-dimensional control vector. The matrices $F_x(i)$ (dimensioned $n \times n$) and

$F_u(i)$ $(n \times m)$ are system matrices that may vary from stage to stage. It is assumed that the initial conditions are specified, i.e.,

$$x(0) = x_0 \qquad \text{specified} \tag{3.30}$$

The quadratic performance index J is chosen to be

$$J = x^{\mathrm{T}}(N)\,\phi_{xx}x(N) + \sum_{i=0}^{N-1} \{x^{\mathrm{T}}(i)\,L_{xx}(i)\,x(i) + u^{\mathrm{T}}(i)\,L_{uu}(i)\,u(i)\} \tag{3.31}$$

where the matrices L_{xx}, L_{uu}, and ϕ_{xx} are dimensioned $n \times n$, $m \times m$, and $n \times n$, respectively. (The more general case, in which linear terms are included, is considered as an exercise.)

It shall be shown by induction that the optimal control law and the optimal return function have the following form,

$$u^{\mathrm{opt}}(x(i)) = -D(i)\,x(i) \tag{3.32}$$

and

$$V^{\mathrm{opt}}(x(i)) = x^{\mathrm{T}}(i)\,P(i)\,x(i) \tag{3.33}$$

where $D(i)$ and $P(i)$ are $m \times n$ and $n \times n$ matrices which will also be determined.

First note that at the final time

$$V^{\mathrm{opt}}(x(N)) = x^{\mathrm{T}}(N)\,\phi_{xx}x(N) \tag{3.34}$$

and, hence, assuming that V^{opt} has the form of Equation (3.33),

$$P(N) = \phi_{xx} \tag{3.35}$$

From Equation (3.25), $u^{\mathrm{opt}}(x(N-1))$ may be written

$$u^{\mathrm{opt}}(x(N-1)) = \arg\max_{u(N-1)}\{R(x(N-1), u(N-1))\} \tag{3.36}$$

where from the definition of R, Equation (3.20),

$$\begin{aligned} &R(x(N-1), u(N-1)) \\ &\quad = x^{\mathrm{T}}(N-1)\,L_{xx}(N-1)\,x(N-1) + u^{\mathrm{T}}(N-1)\,L_{uu}(N-1)\,u(N-1) \\ &\qquad + x(N)^{\mathrm{T}}\,P(N)\,x(N) \end{aligned} \tag{3.37}$$

Next, $x(N)$ is expressed in terms of $x(N-1)$ and $u(N-1)$ by using the system equations (3.29) giving

$$\begin{aligned} &R(x(N-1), u(N-1)) \\ &\quad = x^{\mathrm{T}}(N-1)\,L_{xx}(N-1)\,x(N-1) + u^{\mathrm{T}}(N-1)\,L_{uu}(N-1)\,u(N-1) \\ &\qquad + (F_x(N-1)\,x(N-1) + F_u(N-1)\,u(N-1))^{\mathrm{T}}\,P(N) \\ &\qquad \times (F_x(N-1)\,x(N-1) + F_u(N-1)\,u(N-1)) \end{aligned} \tag{3.38}$$

This equation may be written

$$R(x(N-1), u(N-1)) \\ = x^{\mathrm{T}}(N-1)\, Z_{xx}(N-1)\, x(N-1) \\ + 2u(N-1)\, Z_{ux}(N-1)\, x(N-1) + u^{\mathrm{T}}(N-1)\, Z_{uu}(N-1)\, u(N-1) \tag{3.39}$$

where

$$\begin{aligned} Z_{xx}(N-1) &= L_{xx}(N-1) + F_x^{\mathrm{T}}(N-1)\, P(N) F_x(N-1) \\ Z_{ux}(N-1) &= F_u^{\mathrm{T}}(N-1)\, P(N) F_x(N-1) \\ Z_{uu}(N-1) &= L_{uu}(N-1) + F_u^{\mathrm{T}}(N-1)\, P(N) F_u(N-1) \end{aligned} \tag{3.40}$$

Now performing the indicated maximization of R with respect to u [Equation (3.36)], by differentiating and setting the resultant partial derivative equal to zero, gives

$$\partial R/\partial u = 0 = 2Z_{ux}(N-1)\, x(N-1) + 2Z_{uu}(N-1)\, u(N-1) \tag{3.41}$$

Then solving for $u(N-1)$,

$$u^{\mathrm{opt}}(N-1) = -Z_{uu}^{-1}(N-1)\, Z_{ux}(N-1)\, x(N-1) \tag{3.42}$$

and, hence, (cf Equation 3.32),

$$D(N-1) = Z_{uu}^{-1}(N-1)\, Z_{ux}(N-1) \tag{3.43}$$

Using this value of u^{opt} in the equation (3.26) for $V^{\mathrm{opt}}(x(N-1))$ gives

$$V^{\mathrm{opt}}(x(N-1)) \\ = x(N-1)^{\mathrm{T}} \{Z_{xx}(N-1) - Z_{xu}(N-1)\, Z_{uu}^{-1}(N-1)\, Z_{ux}(N-1)\}\, x(N-1) \tag{3.44}$$

which is of the form of Equation (3.33) if $P(N-1)$ is chosen so that

$$P(N-1) = Z_{xx}(N-1) - Z_{xu}(N-1)\, Z_{uu}^{-1}(N-1)\, Z_{ux}(N-1) \tag{3.45}$$

In the foregoing it is assumed that $Z_{uu}(N-1)$ is nonzero. In fact, in order that the control law equation (3.42) maximizes the return, $Z_{uu}(N-1)$ must be negative definite, i.e., $Z_{uu}(N-1) < 0$.

Now, to complete the induction assume that the optimal control law, $u^{\mathrm{opt}}(x(i))$, and the optimal return, $V^{\mathrm{opt}}(x(i))$, have been obtained for $i = (k+1, \ldots, N-1)$ and that $V^{\mathrm{opt}}(x(k+1))$ has the form

$$V^{\mathrm{opt}}(x(k+1)) = x(k+1)^{\mathrm{T}}\, P(k+1)\, x(k+1) \tag{3.46}$$

The optimal value of u at the kth stage is given by

$$u^{\mathrm{opt}}(x(k)) = \arg\max_{u(k)}\{R(x(k), u(k))\} \tag{3.47}$$

where

$$R(x(k), u(k)) = x^{\mathrm{T}}(k) L_{xx}(k)\, x(k) + u^{\mathrm{T}}(k) L_{uu}(k)\, u(k) + x(k+1)^{\mathrm{T}} P(k+1)\, x(k+1) \tag{3.48}$$

Following the analysis just given for the $N-1$ stage [Equations (3.37)–(3.39)], Equation (3.48) is written

$$R(x(k), u(k)) = x^{\mathrm{T}}(k) Z_{xx}(k)\, x(k) + 2u^{\mathrm{T}}(k) Z_{ux}(k)\, x(k) + u^{\mathrm{T}}(k) Z_{uu}(k)\, u(k) \tag{3.49}$$

where Z_{uu}, Z_{ux}, and Z_{xx} are given by Equation (3.40) with $N-1$ replaced by k. Again maximizing with respect to u gives

$$u^{\mathrm{opt}}(k) = -Z_{uu}^{-1}(k) Z_{ux}(k)\, x(k) \tag{3.50}$$

and, hence,

$$D(k) = Z_{uu}^{-1}(k) Z_{ux}(k) \tag{3.51}$$

Finally, $V^{\mathrm{opt}}(x(k))$ is given by [cf Equation (3.25)]

$$V^{\mathrm{opt}}(x(k)) = x^{\mathrm{T}}(k)\{Z_{xx}(k) - Z_{xu}(k) Z_{uu}^{-1}(k) Z_{ux}(k)\}\, x(k) \tag{3.52}$$

Hence, V^{opt} is the form of Equation (3.33) for all k, provided that $P(k)$ is chosen such that

$$P(k) = Z_{xx}(k) - Z_{xu}(k) Z_{uu}^{-1}(k) Z_{ux}(k) \tag{3.53}$$

Again, for a maximum, the matrix $R_{uu}(k)$ is required to be negative definite for all k, i.e.,

$$Z_{uu}(k) < 0, \qquad k = 0, 1, \ldots, N-1 \tag{3.54}$$

This completes the induction proof. Equations (3.35), (3.51), and (3.53) may be used to compute the optimal control law and the optimal return function in a backwards sequence.

3.4.2 Uniqueness and Sufficiency

In obtaining the dynamic programming solution to this problem the convexity condition $Z_{uu}(k) < 0$ was required. If $Z_{uu}(k)$, for some k, is greater than zero, it is clear that a bounded optimal solution does not exist. Also, if $Z_{uu}(k)$ is singular, the optimal solution will not be unique.

Assuming that neither of the foregoing cases hold, a uniqueness and sufficiency theorem follows.

THEOREM 3.1. If

$$Z_{uu}(k) < 0, \qquad k = 0, 1, \ldots, N-1 \tag{3.55}$$

the optimal control solution of the foregoing problem exists, is unique, and is provided by the above dynamic programming equations.

PROOF. The proof follows directly from the derivation of the dynamic programming equations.

3.4.3 Summary of the Algorithm

The algorithm may be summarized as follows.

1. Set $P(N) = \phi_{xx}$ and $k = N - 1$.
2. Evaluate $Z_{xx}(k)$, $Z_{ux}(k)$, and $Z_{uu}(k)$ from the equations

$$\begin{aligned} Z_{xx}(k) &= L_{xx}(k) + F_x^{\mathrm{T}}(k)\, P(k+1)\, F_x(k) \\ Z_{ux}(k) &= F_u^{\mathrm{T}}(k)\, P(k+1)\, F_x(k) \\ Z_{uu}(k) &= L_{uu}(k) + F_u^{\mathrm{T}}(k)\, P(k+1)\, F_u(k) \end{aligned} \tag{3.56}$$

3. Compute $D(k)$ where

$$D(k) = Z_{uu}^{-1}(k)\, Z_{ux}(k) \tag{3.57}$$

4. Compute $P(k)$ where

$$P(k) = Z_{xx}(k) - Z_{xu}(k)\, Z_{uu}^{-1}(k)\, Z_{ux}(k) \tag{3.58}$$

5. Set $k = k - 1$ and repeat steps 2 and 3 until $k = 0$.
6. Determine $u^{\mathrm{opt}}(k)$ $(k = 0, N - 1)$ from

$$u^{\mathrm{opt}}(k) = -D(k)x(k) \tag{3.59}$$

3.4.4 Example

Low-Thrust Guidance. The sequential algorithm developed in the last section will now be applied to a simple example. Although the numbers involved refer to a low-thrust space probe, similar equations arise in many other problems and the techniques employed are universal.

Consider the space probe traveling along a trajectory from the Earth to Jupiter. Only the motion in a plane normal to the trajectory is of interest as time of flight errors are assumed to be relatively unimportant (Figure 3.1).

The motion of the probe in this plane may be approximated by the equations

$$\ddot{x} = u_x, \qquad \ddot{y} = u_y \tag{3.60}$$ †

where x and y are the components of the position deviations centered on the trajectory, and where u_x and u_y are the control variables in the x and y directions. As motion in the x and y directions is assumed to be uncoupled, only one axis will be considered. The problem is, given some

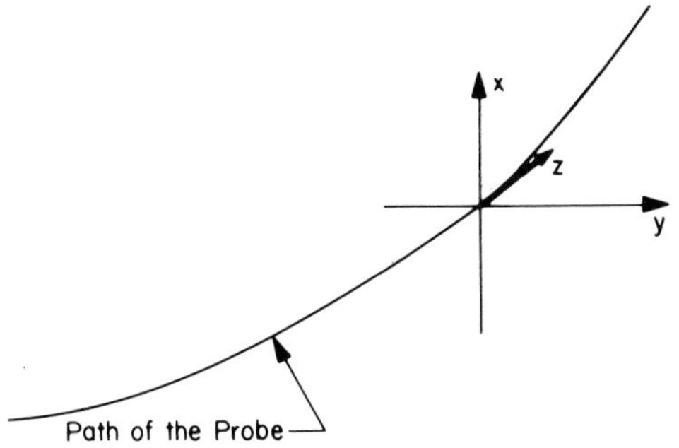

FIG. 3.1. Definition of probe coordinate system.

initial displacement, to minimize the deviation in position x and velocity $\dot{x}$ at the final time (which is given), while minimizing the control u_x. In vector notation the system equations are written,

$$\begin{aligned} \dot{x}_1 &= au, \qquad & x_1(0) &= x_1^0 \\ \dot{x}_2 &= x_1, \qquad & x_2(0) &= x_2^0 \end{aligned} \Big\} \text{ specified} \tag{3.61}$$

where a is a system parameter and x_1 is the velocity deviation (kilometers per second), x_2 is the position deviation (kilometers), and u is the control force (kilometers per square second).

Now, at least two approaches might be taken in the design of a performance index, for example, one criterion might be

$$J_1 = \lambda(x_1(t_f)^2 + x_2(t_f)^2) + \int_0^{t_f} \{\gamma u^2\}\, dt \tag{3.62}$$

where t_f is the final time. The value of the performance index is proportional to the state variables x_1 and x_2 at the final time, and the total control effort used. The parameters λ and γ adjust the relative importance of the

† The overdots denote differentiation with respect to time.

terms in performance index. The other approach might be to include a measure of the state throughout the trajectory, for example,

$$J_2 = \lambda(x_1(t_f)^2 + x_2(t_f)^2) + \int_0^{t_f} \{x_1(t)^2 + x_2(t)^2 + \gamma u^2(t)\}\, dt \tag{3.63}$$

It might be thought that both indices would lead to similar control laws and similar trajectories, but, in general, this is not the case. The difference is well illustrated by this example.

First, however, consider some of the system parameters. For a Jupiter mission, in the mid 1970s, typical flight times are of the order of 900 days of which the last 430 would be unpowered flight where no guidance would be possible. Suppose, for the purpose of this illustration, only the last 50 days of powered flight are considered. Typical errors in position and velocity at the initial (50 days before cutoff) time would be 3000 km and 0.3 m/sec. At this distance from the sun (1–2 AU)† assuming an ion engine, the thrust available for guidance corrections would be equivalent to an acceleration of 10^{-4} km/sec^2. Final position and velocity errors should be of the order of 30 km and 3 mm/sec.

It is obvious from the time scale that continuous control is somewhat unrealistic, as this would require continuous monitoring of the probe's position. A more realistic approach would be to consider the 50 days as four stages,‡ and the final unpowered part of the flight as the final fifth stage. The problem may thus be reformulated as a linear multistage problem. The state transition equations may be obtained by integrating the equations (3.61) over one stage, i.e.,

$$\begin{aligned} x_1(i+1) &= x_1(i) + h(i)\, a(i)\, u(i) \\ x_2(i+1) &= x_2(i) + h(i)\, x_1(i) + \tfrac{1}{2}h(i)^2\, a(i)\, u(i), \qquad i = 0, 1, \ldots, 4 \end{aligned} \tag{3.64}$$

where $h(i)$ is the length of the stage. Hence the matrices F_x and F_u are given by

$$F_x(i) = \begin{bmatrix} 1 & 0 \\ h(i) & 1 \end{bmatrix}, \qquad F_u(i) = \begin{bmatrix} h(i)a(i) \\ \tfrac{1}{2}h^2(i)a(i) \end{bmatrix} \tag{3.65}$$

The performance indices J_1 and J_2 become

$$J_1 = \lambda(x_1(5)^2 + x_2(5)^2) + \sum_{i=0}^{i=4} \{\gamma h(i)\, u(i)^2\} \tag{3.66}$$

† 1 AU (astronomical unit) $\simeq 149.6 \times 10^6$ km.

‡ The choice of the number of stages was made somewhat arbitrarily for this example. In practice many different physical, practical, and economic constraints might have to be considered before making a final decision.

and

$$J_2 = \lambda(x_1(5)^2 + x_2(5)^2) + \sum_{i=0}^{i=4} \{x_1(i)^2 + x_2(i)^2 + \gamma u(i)^2\} h(i) \tag{3.67}$$

Hence the matrices L_{xx}, L_{uu}, and ϕ_{xx} are, for J_1,

$$L_{xx}(i) = \begin{bmatrix} 0 & 0 \\ 0 & 0 \end{bmatrix}, \qquad L_{uu}(i) = [\gamma h(i)], \qquad \phi_{xx} = \begin{bmatrix} \lambda & 0 \\ 0 & \lambda \end{bmatrix} \tag{3.68}$$

and for J_2,

$$L_{xx}(i) = \begin{bmatrix} h(i) & 0 \\ 0 & h(i) \end{bmatrix}, \qquad L_{uu}(i) = [\gamma h(i)], \qquad \phi_{xx} = \begin{bmatrix} \lambda & 0 \\ 0 & \lambda \end{bmatrix} \tag{3.69}$$

One feature of this problem is the large numbers involved. And, as is often the case, it is worthwhile to scale the variables. Here units of time were taken as megaseconds (seconds $\times$ 10^6) and, of distance, megameters (meters $\times$ 10^6). In these units each of the first four stages becomes one unit of time, and the last stage becomes 35 units, i.e.,

$$h(i) = 1, \qquad h(4) = 35, \qquad i = 0, 1, 2, 3$$

Control was possible only over the first four stages, hence,

$$a(i) = 1, \qquad a(4) = 0, \qquad i = 0, 1, 2, 3$$

In these units the thrust available for guidance was approximately one unit, and so the value of the weighting parameter γ had to be chosen to limit the thrust to about this value. Here $\gamma = 100$ was found to be suitable. The value of λ was chosen to ensure that the terminal errors were within the allowed limits, $\lambda = 39$ was found to be suitable. It must be emphasized that the choice of these parameters, γ and λ, is not critical.

The sequential relations equations (3.56) and (3.59) were programmed, and the optimal control computed for both performance indices. The optimal control and the optimal trajectory are given in Tables V and VI and are plotted on Figures 3.2–3.4.

It can be seen that there are significant differences between the two trajectories although the final states are similar. The first performance index J_1 leads to a far more economical fuel consumption although the majority of the trajectory is a long way off the nominal. The second, J_2, leads to a trajectory that follows the nominal far more closely although much more control effort is needed. Of course, the terminal state may not be the only consideration when a trajectory is chosen. On-board experiments may dictate other factors. However, from this simple

TABLE V. Low-Thrust Problem Optimal Trajectory J_1

Stage	x_1	x_2	u
0	0.30000000D–00	0.30000000D–01	−0.78053646D–01
1	0.22194635D–00	0.33000000D–01	−0.77817659D–01
2	0.14412870D–00	0.35219464D–01	−0.77581671D–01
3	0.66547024D–01	0.36660750D–01	−0.77345684D–01
4	−0.10798660D–01	0.37326221D–01	−0.00000000D–38
5	−0.10798660D–01	0.60509528D–03	

Weighted terminal error is 0.45621110D–02.
Fuel used is 0.311.

TABLE VI. Low-Thrust Problem J_2

Stage	x_1	x_2	u
0	0.30000000D–00	0.30000000D–01	−0.11489390D–01
1	−0.84893898D–00	0.33000000D–01	−0.41724174D–00
2	−0.12661807D–01	0.24510610D–01	0.28577248D–00
3	−0.98040824D–00	0.11848803D–00	0.97579561D–00
4	−0.61262721D–03	0.20447207D–00	−0.00000000D–38
5	−0.61262721D–03	−0.72518953D–02	

Weighted terminal error is 0.20656466D–02.
Fuel used is 1.7.

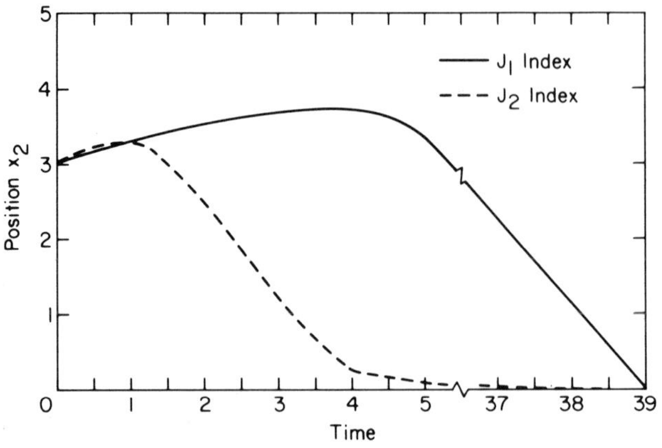

FIG. 3.2. Optimal trajectories (position). —— J_1 index, – – – J_2 index.

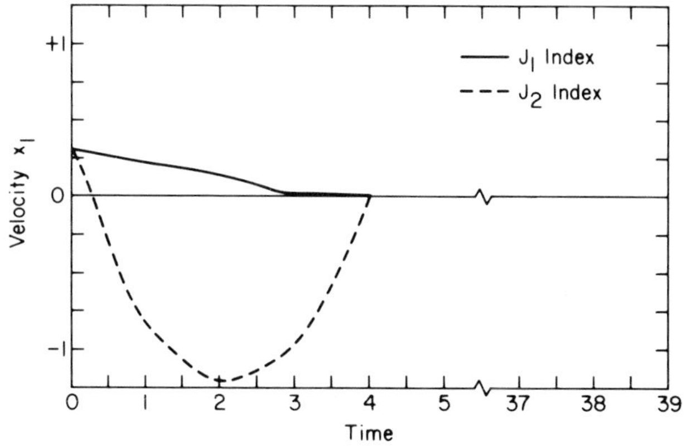

FIG. 3.3. Optimal trajectories (velocity). —— J_1 index, - - - J_2 index.

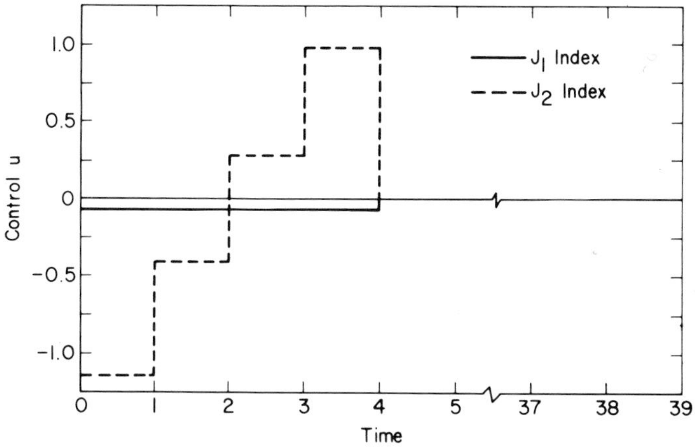

FIG. 3.4. Optimal control histories. —— J_1 index, - - - J_2 index.

analysis it appears that it would be expensive to follow the nominal trajectory closely, and, in terms of the final position, little would be gained.

3.4.5 Linear Quadratic Problems with Linear Terminal Constraints

An analytic solution to the linear quadratic problem also exists if linear terminal constraints are present. These constraints will be denoted by,

$$K = \psi_x x(N) \tag{3.70}$$

where $K = (K_1, K_2, \ldots, K_r)$ denotes the constraint levels. The dimension of K must be no greater than that of n $(r \leqslant n)$. The vector K may be considered as a specified constant, and the solution will be obtained for an arbitrary K. ψ_x is an $r \times n$ matrix and must have full rank, i.e., the constraints must be linearly independent.

This form of terminal constraint is a special case of the general constraint given in Equation (3.6). However, Equation (3.6) may be converted into the terminal constraint by introducing a new state variable x_{n+1} where

$$x_{n+1}(0) = 0$$
$$x_{n+1}(i+1) = x_{n+1}(i) + M(x(i), \alpha, u(i)), \qquad i = 0, 1, \ldots, N-1$$

Thus the general constraint may be rewritten $K = \psi_x(x(N)) + x_{n+1}(N)$. Hence, if M is a linear function of its arguments, the more general form is included.

The first step in deriving the algorithm is to form an augmented performance index J^* by adjoining the terminal constraints to the old index with a set of Lagrange multipliers ν, i.e., form,

$$\begin{aligned} J^* &= x^{\mathrm{T}}(N)\phi_{xx}x(N) + \nu^{\mathrm{T}}(\psi_x x(N) - K) \\ &\quad + \sum_{i=0}^{N-1} \{x^{\mathrm{T}}(i)L_{xx}(i)\,x(i) + u^{\mathrm{T}}(i)L_{uu}(i)\,u(i)\} \end{aligned} \tag{3.71}$$

Now, following the development for the unconstrained problem, an extremal control law and the corresponding return function are sought in terms of x and ν. They are assumed to be of the form

$$\begin{aligned} u^{\mathrm{ext}}(x(k), \nu) &= -D(k)\,x(k) - E(k)\,\nu \\ V^{\mathrm{ext}}(x(k), \nu) &= x^{\mathrm{T}}(k)\,P(k)\,x(k) + 2x^{\mathrm{T}}(k)\,Q(k)\,\nu \\ &\quad + \nu^{\mathrm{T}}S(k)\,\nu - \nu^{\mathrm{T}}K \end{aligned} \tag{3.72}$$

Now, as before, assume that $V^{\mathrm{ext}}(x(k+1), \nu)$ has been obtained, and u^{ext} may be found from

$$u^{\mathrm{ext}}(x(k), \nu) = \arg \operatorname*{ext}_{u(k)}[R(x(k), \nu, u(k))]$$

where

$$\begin{aligned} R(x(k), \nu, u(k)) &= x^{\mathrm{T}}(k)L_{xx}(k)\,x(k) + u^{\mathrm{T}}(k)L_{uu}(k)\,u(k) \\ &\quad + x^{\mathrm{T}}(k+1)\,P(k)\,x(k+1) \\ &\quad + 2x^{\mathrm{T}}(k+1)\,Q(k+1)\,\nu + \nu^{\mathrm{T}}S(k+1)\,\nu - \nu^{\mathrm{T}}K \end{aligned} \tag{3.73}$$

and where

$$x(k+1) = F_x(k)\,x(k) + F_u(k)\,u(k)$$

This expression, Equation (3.73), is, for convenience, rewritten as

$$\begin{aligned} R(x(k), \nu, u(k)) = {} & x^{\mathrm{T}}(k)\,Z_{xx}(k)\,x(k) \\ & + 2u^{\mathrm{T}}(k)\,Z_{ux}(k)\,x(k) + u^{\mathrm{T}}(k)\,Z_{uu}(k)\,u(k) \\ & + 2x^{\mathrm{T}}(k)\,Z_{x\nu}(k)\,\nu + 2u^{\mathrm{T}}(k)\,Z_{u\nu}(k)\,\nu \\ & + \nu^{\mathrm{T}} Z_{\nu\nu}(k)\,\nu - \nu^{\mathrm{T}} K \end{aligned} \tag{3.74}$$

where Z_{xx}, Z_{ux}, and Z_{uu} are defined as before by Equation (3.40), and

$$Z_{x\nu}(k) = F_x^{\mathrm{T}}(k)\,Q(k+1), \qquad Z_{\nu\nu}(k) = S(k+1)$$

The extremal control $u^{\mathrm{ext}}(k)$ is easily found by differentiating this expression with respect to $u(k)$ and setting to zero, giving

$$u^{\mathrm{ext}}(k) = -Z_{uu}^{-1}(k)\,Z_{ux}(k)\,x(k) - Z_{uu}^{-1}(k)\,Z_{u\nu}(k)\,\nu \tag{3.75}$$

Hence,

$$D(k) = Z_{uu}^{-1}(k)\,Z_{ux}(k), \qquad E(k) = Z_{uu}^{-1}(k)\,Z_{u\nu}(k) \tag{3.76}$$

Now, by definition,

$$V^{\mathrm{ext}}(x(k), \nu) = R(x(k), \nu, u^{\mathrm{ext}}(x(k), \nu))$$

and so by substituting $u^{\mathrm{ext}}(k)$ [Equation (3.75)] back into Equation (3.74), it may easily be seen that $V^{\mathrm{ext}}(x(k), \nu)$ will have the quadratic form of Equation (3.72), provided P, Q, and S are chosen by

$$\begin{aligned} P(k) &= Z_{xx}(k) - Z_{xu}(k)\,Z_{uu}^{-1}(k)\,Z_{ux}(k) \\ Q(k) &= Z_{x\nu}(k) - Z_{xu}(k)\,Z_{uu}^{-1}(k)\,Z_{u\nu}(k) \\ S(k) &= Z_{\nu\nu}(k) - Z_{\nu u}(k)\,Z_{uu}^{-1}(k)\,Z_{u\nu}(k) \end{aligned} \tag{3.77}$$

These relations, Equations (3.75)–(3.77), enable V^{ext} and u^{ext} to be computed as a backwards sequence. The relations are initialized by noting that

$$V^{\mathrm{ext}}(x(N), \nu) = x^{\mathrm{T}}(N)\,\phi_{xx} x(N) + (x^{\mathrm{T}}(N)\,\psi_x{}^{\mathrm{T}} - K)\,\nu$$

which implies that $P(N) = \phi_{xx}$, $Q(N) = \psi_x{}^{\mathrm{T}}$, $S(N) = 0$.

Note that these relations have been obtained for an arbitrary value of ν. The multiplier ν must be chosen before determining whether the solution maximizes the performance index. This is done by extremizing the return $V^{\text{ext}}(x(k), \nu)$ with respect to ν, i.e.,

$$\nu^{\text{ext}} = \arg\operatorname*{ext}_{\nu}\{V^{\text{ext}}(x(k), \nu)\} \tag{3.78}$$

Equation (3.78) leads to $\nu^{\text{ext}} = S^{-1}(k)(K - Q^{T}(k)\, x(k))$ providing the matrix S can be inverted. Although in this equation ν^{ext} appears to vary from stage to stage, it is, in fact, a constant over a specific optimal trajectory. In other words, the multiplier ν may be found at any point along a trajectory so long as the matrix S is nonsingular. Substituting this expression for ν into the equations for u^{ext} and V^{ext}, the optimal control law and the optimal return are obtained, viz,

$$u^{\text{opt}}(x(k), K) = -D(k) - E(k)\, S^{-1}(k)\, Q^{\text{T}}(k)\, x(k) - E(k)\, S^{-1}(k)\, K$$

and

$$\begin{aligned} V^{\text{opt}}(x(k), K) &= x^{\text{T}}(k)(P(k) - Q^{\text{T}}(k)\, S^{-1}(k)\, Q(k))\, x(k) \\ &\quad - x^{\text{T}}(k)\, Q(k)\, S^{-1}(k)\, K + K^{\text{T}} S^{-1}(k)\, K \end{aligned} \tag{3.79}$$

As noted before this construction requires that $S(k)$ has full rank. If, for some k, $S(k)$ is singular, ν^{ext} must be computed at a previous stage and the resulting control is essentially open loop.

If at the jth stage S obtains full rank, then the algorithm may be simplified by computing V^{opt} and u^{opt} for the preceeding stages explicitly. In such cases, assuming that $K = 0$, $V^{\text{opt}}(x(k))$ and $u^{\text{opt}}(x(k))$ have the following forms,

$$\left.\begin{aligned} V^{\text{opt}}(x(k)) &= x^{\text{T}}(k)\, P^*(k)\, x(k), \\ u^{\text{opt}}(x(k)) &= -D^*(k)\, x(k), \end{aligned}\right\} \quad k = j, j-1, \ldots, 0$$

where $P^*(k)$ and $D^*(k)$ satisfy the same relations as $P(k)$ and $D(k)$ for the unconstrained problem. However, the initialization for $P^*(k)$ is provided by Equation (3.79) with the subscript k replaced by j.

3.5 NUMERICAL METHODS

Unfortunately, in general elegant analytical solutions to the dynamic programming equations, such as that obtained for the linear quadratic problem, are not possible. Instead numerical techniques must be employed. The traditional dynamic programming approach involves a systematic search procedure that is best described in terms of a simple example.

Consider a very simple one-dimensional regulator problem. The system equation is $x(k+1) = x(k) + u(k)$ where k equals 0, 1, ..., and the performance index, to be minimized, is

$$J = \sum_{k=0}^{1} \{x^2(k) + u^2(k)\} + x(2)^2$$

The conventional approach would be as follows.

1. Compute values of the optimal return $V^{\text{opt}}(x(2))$ for a range of values of $x(2)$, e.g., for ten values in the range $0 \leqslant x(2) \leqslant 1$, and construct a table of $V^{\text{opt}}(x(2))$ versus $x(2)$. (See Table VII.)

TABLE VII. $V^{\text{opt}}(x(2))$

V^{opt}	0	0.01	0.04	0.09	0.16	0.25	0.36	0.49	0.64	0.81	1
$x(2)$	0	0.1	0.2	0.3	0.4	0.5	0.6	0.7	0.8	0.9	1

2. The next step is to compute a similar table for values of $V^{\text{opt}}(x(1))$ vs. $x(1)$. The optimal value of $u(1)$ must maximize the return $R(x(1), u(1))$ where

$$R(x(1), u(1)) = x(1)^2 + u(1)^2 + V^{\text{opt}}(x(2))$$

with $x(2)$ given by $x(2) = x(1) + u(1)$. Again specific values of $x(1)$ for $0 < x(1) < 1$ are considered. The control $u(1)$, in some algorithms, may only take discrete values, whereas in others is not so constrained. Consider first $x(1) = 0$, then $R(0, u(1)) = u(1)^2 + V^{\text{opt}}(u(1))$ which is clearly a minimum if $u(1) = 0$. Next, consider $x(1) = 0.1$,

$$R(0.1, u(1)) = 0.01 + u(1)^2 + V^{\text{opt}}(0.1 + u(1))$$

We might try values for u in the range $-0.1 \leqslant u \leqslant 0.1$. Note that values for $V^{\text{opt}}(0.1 + u(1))$ have to be interpolated if $0.1 + u(1)$ does not equal one of the stored values of $x(2)$. Values of R for values of $u(1)$ are given in Table VIII where it can be seen that $u(1) = -0.05$ gives the minimum. The procedure is repeated for each value of $x(1)$ so that a table of values of $V^{\text{opt}}(x(1))$, $x(1)$ and $u^{\text{opt}}(1)$ may be constructed.

TABLE VIII. $R(0.1, u(1))$

$R(0.1, u(1))$	0.02	0.0075	0.02	0.035	0.06
$u(1)$	-0.1	-0.05	0	0.05	0.1

3. The last step is repeated for $x(0)$ and $u(0)$ until finally a table of $V^{\text{opt}}(x(0))$, $x(0)$ and $u^{\text{opt}}(0)$ is constructed.

The general procedure may be summarized as follows. The state and the control are divided into a number of discrete values. The value of the optimal return $V^{\text{opt}}(x(N))$ is computed simply by evaluating $\phi(x(N))$ for each of the quantized states. Then $V^{\text{opt}}(x(N-1))$ is computed by evaluating $R(x(N-1), u(N-1))$ for each control value and the maximum is found by an elementary trial and error procedure. It may, of course, be necessary to interpolate the values of $V^{\text{opt}}(x(N))$ if $x(N) = F(x(N-1), u(N-1))$ is not a quantized state. The procedure is repeated for each value of $x(N-1)$ so that a table may be constructed of V^{opt}, $x(N-1)$, and $u^{\text{opt}}(N-1)$. This procedure is repeated at each stage until the initial stage is reached. The final result is a table of values of $V^{\text{opt}}(x(0))$, $x(0)$, and $u^{\text{opt}}(0)$.

There are certain advantages to this procedure. For example, very general systems can be handled, and constraints actually tend to help the computation by reducing the search areas. Also the optimal control is obtained as a feedback solution. In practice, however, the computational requirements become excessive for all but very simple systems. Both storage requirements and computation time can be troublesome. For example, most of the storage is required for the table of V^{opt} as a function of the state. For the n-dimensional case with 100 quantized values of the state, 100^n storage locations would be required.

As a consequence several ingenious methods have been developed that attempt to reduce the amount of storage required, e.g., if there are fewer control variables m than state variables n, a substantial saving may be achieved by transforming the state space.

A completely different approach is to obtain the optimal solution from a single initial state. The procedures developed here are iterative in nature, i.e., they are based on updating some nominal control sequence such that the optimum is eventually reached. The storage and time requirements are comparatively small. A true feedback solution is not obtained, although there exist techniques by which it is possible to compute the optimum feedback in a small region about the optimal trajectory.

3.6 THE GRADIENT METHOD

3.6.1 The Gradient Algorithm

In this section a gradient method for the optimization of multistage problems will be developed. The method is motivated from a consideration of a first-order expansion of the return function, $V(x(0), \alpha, u[0, N-1])$, about some nominal control variable sequence $u^j[0, N-1]$ and a nominal parameter, α^j, viz,

$$V(x(0), \alpha^{j+1}, u^{j+1}[0, N-1]) = V(x(0), \alpha^j, u^j[0, N-1]) + \frac{\partial V(x(0), \alpha^j, u^j[0, N-1])}{\partial \alpha} \delta\alpha + \frac{\partial V(x(0), \alpha^j, u^j[0, N-1])}{\partial u[0, N-1]} \delta u[0, N-1] \tag{3.80}$$

The variations in the control variables $\delta\alpha = \alpha^{j+1} - \alpha^j$ and $\delta u = u^{j+1} - u^j$ must be small enough to ensure the validity of the expansion. Clearly, if $\delta u[0, N-1]$ and $\delta\alpha$ are chosen by

$$\delta u[0, N-1] = \epsilon \left[\frac{\partial V(x(0), \alpha^j, u^j[0, N-1])}{\partial u[0, N-1]}\right]^{\mathrm{T}} \tag{3.81}$$

and

$$\delta\alpha = \epsilon \left[\frac{\partial V(x(0), \alpha^j, u^j[0, N-1])}{\partial \alpha}\right]^{\mathrm{T}} \tag{3.82}$$

where ϵ is some positive parameter, the return

$$V(x(0), \alpha^{j+1}, u^{j+1}[0, N-1])$$

will be greater than $V(x(0), \alpha^j, u^j[0, N-1])$.

Instead of forming $V(x(0), \alpha, u[0, N-1])$ explicitly and then differentiating, the gradients are computed more easily by means of a backward sequence of equations.

It is clear that a change in $u(k)$ will not affect $L(x(i), \alpha, u(i))$ for $i < k$. Hence, the gradient of the return function $V(x(0), \alpha, u[0, N-1])$ with respect to the control function $u(k)$ is the same as the gradient of $V(x(k), \alpha, u[k, N-1])$, i.e.,

$$\frac{\partial V(x(0), \alpha, u[0, N-1])}{\partial u(k)} = \frac{\partial V(x(k), \alpha, u[k, N-1])}{\partial u(k)} \tag{3.83}$$†

Now the return $V(x(k), \alpha, u[k, N-1])$ from its definition may be written

$$V(x(k), \alpha, u[k, N-1]) = L(x(k), \alpha, u(k)) + V(x(k+1), \alpha, u[k+1, N-1])$$

† The superscript j, denoting the nominal, is dropped to clarify the text.

Hence, differentiating with respect to $u(k)$, we obtain

$$\frac{\partial V(x(k), \alpha, u[k, N-1])}{\partial u(k)}$$

$$= \frac{\partial L(x(k), \alpha, u(k))}{\partial u(k)} + \frac{\partial V(x(k+1), \alpha, u[k+1, N-1])}{\partial x(k+1)} \frac{\partial F(x(k), \alpha, u(k))}{\partial u(k)} \tag{3.84}$$

In the following analysis the arguments $x(i)$, α, $u[i, N-1]$ or $x(i)$, α, $u(i)$ will be replaced by the single argument i. Subscripts, unless specifically defined otherwise, denote partial derivatives. For example, Equation (3.84) is written $V_u(k) = L_u(k) + V_x(k+1) F_u(k)$. In order to evaluate this expression it is necessary to obtain $V_x(k+1)$ for $k = 0, 1, \dots, N-1$. A sequential set of relations for V_x evaluated along a trajectory may be obtained by taking partial derivatives of Equations (3.16) and (3.17), i.e.,

$$V_x(N) = \phi_x(x(N), \alpha) \tag{3.85}$$

$$V_x(k) = V_x(k+1) F_x(k) + L_x(k) \tag{3.86}$$

The partial derivative of $V(0)$ with respect to α is obtained in a similar fashion by partially differentiating Equation (3.16), viz,

$$V_\alpha(N) = \phi_\alpha(x(N), \alpha) \tag{3.87}$$

$$V_\alpha(k) = V_\alpha(k+1) + L_\alpha(k) + V_x(k+1) F_\alpha(k) \tag{3.88}$$

where $V_x(k+1)$ is given by Equations (3.85) and (3.86). Thus using Equations (3.85)–(3.88), the gradients of the return $V(x(0), \alpha, u[0, N-1])$ with respect to α, and $u[0, N-1]$ may be formed as a backward sequence.

The gradient algorithm may now be summarized as follows.

1. Choose a nominal control sequence $u^j[0, N-1]$ and control parameter α^j. Construct and store the trajectory $x[0, N]$ from the system equations (3.4) and the nominal control variables. Also compute the cost J where

$$J = \phi(x(N), \alpha) + \sum_{i=0}^{N-1} \{L(x(i), \alpha, u(i))\} \tag{3.89}$$

2. Compute the partial derivatives $V_x(k) = V_x(x(k), \alpha^j, u^j[k, N-1])$

and $V_\alpha^j(k) = V_\alpha(x(k), \alpha^j, u^j[k, N-1])$ for $k = N, N-1, \dots, 0$ from

$$V_x^j(N) = \phi_x(N), \qquad V_\alpha^j(N) = \phi_\alpha(N)$$
$$V_x^j(k) = V_x^j(k+1) F_x^j(k) + L_x^j(k) \tag{3.90}$$
$$V_\alpha^j(k) = V_x^j(k+1) F_\alpha^j(k) + V_\alpha^j(k+1) + L_\alpha^j(k)$$

where $F_\alpha^j(k) = F_\alpha(x(k), \alpha^j, u^j(k))$ and $L_\alpha^j(k) = L_\alpha(x(k), \alpha^j, u^j(k))$.

3. Compute and store the gradient with respect to the control at each stage from

$$\frac{\partial V(x(0), \alpha^j, u^j[0, N-1])}{\partial u^j(k)} = V_x^j(k+1) F_u^j(k) + L_u^j(k), \qquad k = N-1, \dots, 0 \tag{3.91}$$

4. For some nominal parameter $\epsilon > 0$, compute the new control from

$$u^{j+1}(k) = u^j(k) + \epsilon[\partial V / \partial u^j(k)]^{\mathrm{T}} \tag{3.92}$$

5. At the *initial* stage compute the new control parameter α^{j+1} from

$$\alpha^{j+1} = \alpha^j + \bar{\epsilon} V_\alpha^{\mathrm{T}}(0) \tag{3.92a}$$

where $\bar{\epsilon} > 0$.

6. Use the new control variables $u^{j+1}[0, N-1]$, α^{j+1} and the system equations to construct a new trajectory $x^{j+1}[0, N]$ and compute the new cost J^{j+1}.

7. If $J^{j+1} > J^j$, continue with steps 2 and 3. If $J^{j+1} < J^j$, reduce the step size parameters ϵ, $\bar{\epsilon}$; for example, set $\epsilon = \epsilon/2$, $\bar{\epsilon} = \bar{\epsilon}/2$, and repeat steps 4, 5, and 6, etc.

8. The iterations are continued until either no further increase in the cost is possible, or until the desired accuracy is attained.

If side constraints are present, appropriate modifications must be made. For example, consider a side constraint of the form, [cf Equation (3.6)],

$$K = \psi(x(N), \alpha) + \sum_{i=0}^{N-1} M(x(i), \alpha, u(i)) \tag{3.93}$$

The gradient projection approach taken here is derived by introducing $W(x(k), \alpha, u[k, N-1])$ associated with the constraint so that

$$W(x(k), \alpha, u[k, N-1]) = \psi(x(N), \alpha) + \sum_{i=k}^{N-1} \{M(x(i), \alpha, u(i))\} \tag{3.94}$$

Following a development similar to that just given above for the return V it is possible to derive the partial derivatives W_x, W_α, i.e.,

$$\begin{aligned} W_x(N) &= \psi_x(x(N), \alpha) \\ W_\alpha(N) &= \psi_\alpha(x(N), \alpha) \\ W_x(k) &= W_x(k+1) F_x(k) + M_x(k) \\ W_\alpha(k) &= W_\alpha(k+1) + W_x(k+1) F_\alpha(k) + M_\alpha(k) \end{aligned} \tag{3.95}$$

where the arguments of the return W have again been abbreviated.

Thus, it is possible to form the gradient of W with respect to the control $u(k)$ as

$$\frac{\partial W(x(0), \alpha, u[0, N-1])}{\partial u(k)} = W_x(k+1) F_u(k) + M_u(k) \tag{3.96}$$

The corrections in the gradient algorithm become

$$u^{j+1}(k) = u^j(k) + \epsilon \left[\frac{\partial V(k)}{\partial u(k)} + \nu^{\mathrm{T}} \frac{\partial W(k)}{\partial u(k)}\right]^{\mathrm{T}} \tag{3.97}$$

and

$$a^{j+1} = \alpha^j + \bar{\epsilon} \left[\frac{\partial V(0)}{\partial \alpha} + \nu^{\mathrm{T}} \frac{\partial W(0)}{\partial \alpha}\right]^{\mathrm{T}}$$

It remains to choose values for the parameter ν. They are chosen to enforce a *desired* change δK in the level of the linearized constraint. In terms of the return W, δK is given by

$$\delta K = W_\alpha(0)\, \delta\alpha + W_u(0)\, \delta u[0, N-1].$$

Substituting for $\delta\alpha = \alpha^{j+1} - \alpha^j$ and

$$\delta u[0, N-1] = u^{j+1}[0, N-1] - u^j[0, N-1]$$

gives

$$\begin{aligned} \delta K = {} & \bar{\epsilon} W_\alpha(0)(V_\alpha(0) + \nu^{\mathrm{T}} W_\alpha(0))^{\mathrm{T}} \\ & + \epsilon \sum_{i=0}^{N-1} \{(W_x(i+1) F_u(i) + M_u(i))(V_x(i+1) F_u(i) \\ & + L_u(i))^{\mathrm{T}} + \| W_x(i+1) F_u(i) + M_u(i)\|^2 \nu\} \end{aligned}$$

Hence, ν may be found to be

$$\nu = \left[\bar{\epsilon} \| W_\alpha(0)\|^2 + \epsilon \sum_{i=0}^{N-1} \| W_x(i+1) F_u(i) + M_u(i)\|^2\right]^{-1}$$

$$\times \left[\delta K - \epsilon \sum_{i=0}^{N-1} (W_x(i+1) F_u(i) \right.$$

$$\left. + M_u(i))(V_x(i+1) F_u(i) + L_u(i))^{\mathrm{T}} - W_\alpha(0) V_\alpha(0)^{\mathrm{T}}\right] \tag{3.98}$$

Of course it must be assumed that abnormality does not occur, i.e.,

$$\bar{\epsilon} \| W_\alpha(0)\|^2 + \epsilon \sum_{t=0}^{N-1} \| W_x(i+1) F_u(i) + M_u(i)\|^2 > 0$$

3.6.2 The Penalty Function Technique

Another approach to the problem of fixed terminal constraints is provided by the penalty function technique. A strictly convex function of the constraints is adjoined to the original performance index J, i.e., $J^* = J + C[\psi(x(N), \alpha) - K]^{2r}$ where the parameter r is often taken as unity and where (in a maximization problem) C is a negative constant. The problem is now normal and unconstrained and the straightforward gradient technique may be used. By increasing the magnitude of the parameter C, the term $[\psi(x(N), \alpha) - K]^{2r}$ can be made arbitrarily small.

This approach is appealing from an engineering viewpoint as, in normal circumstances, hard constraints do not have to be met exactly. One disadvantage is that additional extrema may be introduced and convergence to the wrong extremal is possible. Fortunately, however, such cases seem to occur in theory rather than in practice. Another disadvantage is that the addition of penalty functions with large weighting constants C tends to slow convergence. The effect, obviously, depends on the problem, but it can become a major factor in the number of iterations required to obtain a solution.

3.6.3 Example

The Brachistochrone Problem. A classic problem in the calculus of variations is the brachistochrone problem, which may be phrased as follows. Find the path by which a bead may slide down a frictionless wire between two specified points in minimal time. The initial velocity of the bead will be assumed to be specified by $V_0 \neq 0$, and the gravitational field has a constant acceleration, g per unit mass. Restricting the bead to a plane, the position of the particle may be described with the rectangular coordinates (y, z), where z is the verticle component and y is the horizontal component of the position vector. (See Figure 3.5). Assuming that along the optimal path, z may be expressed as a piecewise differential function of y, the time of transfer may be expressed by the path integral,

$$T = \int_{y_0}^{y_n} \frac{[1 + (dz/dy)^2]^{1/2}}{V(z, y)} \, dy$$

where y_0 and y_n are the initial and final values of y and $V(z, y)$ is the velocity of the bead.

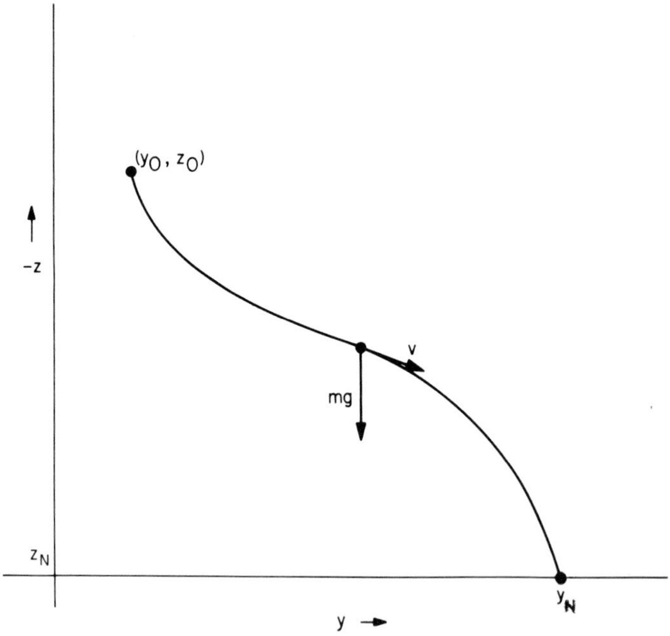

FIG. 3.5. The brachistochrone problem.

The conservation of energy implies $V_0^2/2 + gz_0 = V^2/2 + gz$. Hence, V may be expressed as terms of z and known parameters $V = (V_0 + 2g(z_0 - z))^{1/2}$.

To simplify the form of the performance index, dimensionless variables are introduced

$$x = (2g/V_0^2)(z - z_0), \qquad t = (2g/V_0^2)(y - y_0)$$
$$J = 2gT/V_0, \qquad u = dz/dy$$

The performance index now becomes

$$J = \int_0^{t_N} (1 + u^2)^{1/2}/(1 - x)^{1/2}\, dt$$

The system equation relating the *time* derivative of the *state* vector x to *control* u is simply $dx/dt = u$. The value of the state at the initial time $t = 0$ is $x(0) = 0$. To complete the problem, let the final time t_N and final state x_N be specified by $t_N = 1$ and $x_N = -0.5$.

To apply the discrete gradient method to this problem, the performance index may be replaced with the following discrete approximation,

$$J = \sum_{i=0}^{N-1} \{[1 + u(i)^2]^{1/2}/[1 - x(i)]^{1/2}\}\, h(i)$$

where $h(i)$ is the length of the ith stage. The system equation is approximated by $x(i + 1) = x(i) + h(i)\, u(i)$. In this example 100 equal stages were chosen so that $h(i) = 0.01$, $i = 0, 1, \ldots, 99$, and $N = 100$.

The boundary conditions, Equations (3.90), for the partial derivatives are $V_x(N) = 0$ and $W_x(N) = 1$. The partial derivatives required to compute Equations (3.90) are

$$\begin{aligned} L_x(i) &= \tfrac{1}{2} h(i)(1 + u(i)^2)^{1/2}\,(1 - x(i))^{-3/2} \\ L_u(i) &= h(i)\, u(i)(1 + u(i)^2)^{-1/2}\,(1 - x(i))^{-1/2} \\ M_u(i) = 0, \quad F_x(i) &= 1, \quad \text{and} \quad F_u(i) = h(i) \end{aligned} \tag{3.99}$$

Finally, the nominal value for the control u^j was chosen as

$$\begin{aligned} u(i) &= 0, & i &= 0, 1, \ldots, 49 \\ u(i) &= -0.5, & i &= 50, 51, \ldots, 99 \end{aligned}$$

and the parameter ϵ was chosen as 0.5.

The gradient algorithm was used to determine the optimal control as follows.

1. The nominal u^j was used to compute a trajectory $x[0, N]$ and the nominal cost J^j.

2. The partial derivatives $V_x(k)$, $W_x(k)$ were computed using Equations (3.90) and (3.95).

3. A value was chosen for δK, here, $\delta K = x^{j-1}(N) + 0.5$ and the Lagrange multiplier ν was computed.

4. The gradients $\partial V/\partial u(k)$, $\partial W/\partial u(k)$ were formed using Equations (3.91) and (3.96) for $k = N - 1, \ldots, 0$.

5. With $\epsilon = 0.5$, the new control u^{j+1} was computed [Equation (3.97)].

6. The new control was used to compute a new trajectory and the new cost J^{j+1} was computed.

7. The new cost was checked to ensure $J^{j+1} < J^j$ and steps 2–6 were repeated for 10 iterations.

The results of the computations are shown in Table IX and Figure 3.6. The computer time required was under one second on the IBM 7090 computer.

TABLE IX. BRACHISTOCHRONE PROBLEM

Iteration	Cost	$u(1)$	$x(101)$	ν
Nominal	1.1369744	0.0000000	−0.4999999	−0.0000000
1	1.0251133	−1.1508888	−0.4999997	0.0693148
2	1.0011187	−0.7745588	−0.5000000	0.1827865
3	0.9997580	−0.7875542	−0.4999998	0.2224539
4	0.9995843	−0.7767487	−0.4999998	0.2231138
5	0.9995633	−0.7816071	−0.4999998	0.2228842
6	0.9995604	−0.7799865	−0.4999998	0.2229772
7	0.9995600	−0.7805840	−0.4999998	0.2229454
8	0.9995599	−0.7803639	−0.4999998	0.2229454
9	0.9995599	−0.7804477	−0.4999999	0.2229574

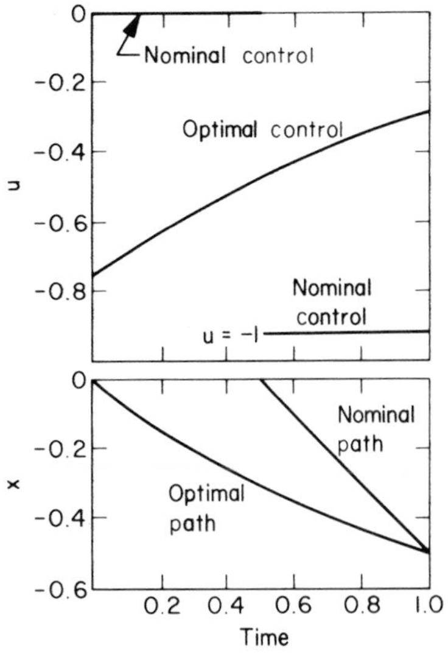

FIG. 3.6. Optimal solution to the brachistochrone problem.

The computation was repeated with the penalty function technique. The terminal constraint was adjoined to the performance index J, i.e., $J^* = J + c(x(N) + 0.5)^2$. The boundary condition for the function V_x becomes $V_x(N) = 2c(x(N) + 0.5)$ and the other partial derivatives remain unaltered. [The partial $W_x(k)$ is, of course, no longer needed.] The gradient algorithm was then used to compute the optimal control for several values of the parameter c.

It was found that the solutions were near to those obtained previously, in the sense that the terminal constraint was met and the value of the returns were similar. However, the control histories were substantially different; in other words, final convergence was not obtained. (See Figure 3.7.)

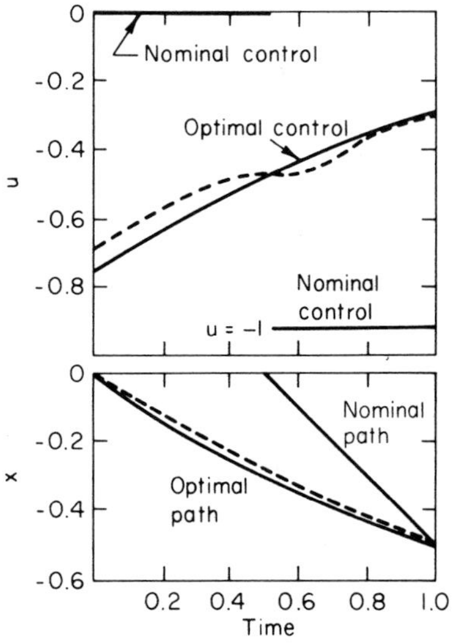

FIG. 3.7. The penalty function solution.

3.6.4 The Stationarity Condition for a Local Maximum

From a consideration of the first-order expansion of the return function equation (3.80) it is clear that the stationarity conditions for the unconstrained problem are

$$\frac{\partial V(x(0), \alpha, u[0, N-1])}{\partial u[0, N-1]} = 0 \tag{3.100}$$

and

$$\frac{\partial V(x(0), \alpha, u[0, N-1])}{\partial \alpha} = 0 \tag{3.101}$$

From the preceding analysis, the first condition may be rewritten

$$V_x(k+1)F_u(k) + L_u(k) = 0, \qquad k = 0, 1, \ldots, N-1 \tag{3.102}$$

The same conditions may also be derived via the classical techniques. The system equations are considered as constraints and adjoined to the performance index with a set of Lagrange multipliers p (a row vector). The augmented performance index J^* is given by

$$J^* = \phi(x(N), \alpha) + \sum_{i-0}^{N-1} [L(x(i), \alpha, u(i))]$$
$$+ \sum_{i=0}^{N-1} \{p(i+1)[F(x(i), \alpha, u(i)) - x(i+1)]\} \tag{3.103}$$

Introducing the notation

$$H(i) = L(x(i), \alpha, u(i)) + p(i+1)F(x(i), \alpha, u(i)) \tag{3.104}$$

Equation (3.103) becomes

$$J^* = \phi(x(N), \alpha) + \sum_{i=0}^{N-1} [H(i) - p(i+1)\,x(i+1)]$$

or alternatively

$$J^* = \phi(x(N), \alpha) - p(N)\,x(N) + \sum_{i=0}^{N-1} [H(i) - p(i)\,x(i)] + p(0)\,x(0) \tag{3.105}$$

Now consider the first variation of J^*, $\delta^1 J^*$, which is given by

$$\delta^1 J^* = [\phi_x(x(N), \alpha) - p(N)]\,\delta x(N) + \sum_{i=0}^{N-1} [H_x(i) - p(i)]\,\delta x(i)$$
$$+ \phi_\alpha(x(N), \alpha)\,\delta\alpha + \sum_{i=0}^{N-1} [H_\alpha(i)]\,\delta\alpha$$
$$+ \sum_{i=0}^{N-1} \{H_u(i)\}\,\delta u(i) + p(0)\,\delta x(0)$$

where the partial derivatives of $H(i)$, $H_x(i)$, $H_u(i)$, and $H_\alpha(i)$ are

$$H_x(i) = L_x(i) + p(i+1)F_x(i)$$
$$H_u(i) = L_u(i) + p(i+1)F_u(i)$$
$$H_\alpha(i) = L_\alpha(i) + p(i+1)F_\alpha(i)$$

So far the Lagrange multipliers p have been considered as arbitrary parameters. They are now chosen by

$$p(N) = \phi_x(x(N), \alpha), \qquad p(i) = H_x(i), \qquad i = 0, 1, \ldots, N-1 \tag{3.106}$$

so that the first variation becomes

$$\begin{aligned} \delta^1 J^* &= p(0)\,\delta x(0) + \sum_{i=0}^{N-1} \{H_u(i)\,\delta u(i) + H_\alpha(i)\,\delta\alpha\} \\ &\quad + \phi_\alpha(x(N), \alpha)\,\delta\alpha \end{aligned} \tag{3.107}$$

But at a local maximum the first variation, δJ^*, must be zero for arbitrary variations in $\delta u(i)$ and $\delta\alpha(0)$. It follows that

$$H_u(i) = 0, \qquad i = 0, 1, \ldots, N-1 \tag{3.108}$$

$$\phi_\alpha(x(N), \alpha) + \sum_{i=0}^{N-1} H_\alpha(i) = 0 \tag{3.109}$$

Equation (3.108) is the same condition as Equation (3.102) if $p(i+1)$ is identified with $V_x(x(i+1), u(i+1), N-1))$, and Equation (3.109) is identical to Equation (3.101).

If side constraints of the form given by Equation (3.6) are present, then the condition for stationarity requires that there exists a Lagrange multiplier ν such that

$$\begin{aligned} V_x(i+1) F_u(i) + L_u(i) + \nu^{\mathrm{T}}(W_x(i) F_u(i) + M_u(i)) &= 0 \\ V_\alpha(0) + \nu^{\mathrm{T}} W_\alpha(0) &= 0 \end{aligned} \tag{3.110}$$

3.6.5 The Discrete Maximum Principle

The function $H(i)$ is often referred to as the Hamiltonian for discrete systems. From Equation (3.108) it may appear that the optimal control maximizes the discrete Hamiltonian $H(i)$; i.e., for the control $u^*[0, N-1]$ to be optimal,

$$H(x(i), u^*(i), p(i+1)) \geqslant H(x(i), u(i), p(i+1)), \qquad i = 0, 1, \ldots, N-1$$

Thus, for a *local* optimum the stationarity condition is $H_u(i) = 0$ and the convexity condition is $H_{uu}(i) \leqslant 0$.

Although it is true that this stationarity condition must hold [cf Equations (3.102) and (3.109)], the convexity condition $H_{uu}(i) \leqslant 0$ *does not, in general, ensure a local optimum.* For some specific types of problems the discrete maximum principle has been shown to hold, e.g.,

Halkin (1966). However the correct convexity condition for the general case will be derived in 3.7.4.

3.7 THE NEWTON–RAPHSON METHOD

3.7.1 The Successive Sweep Algorithm

The successive sweep algorithm is motivated from a consideration of a second-order expansion of the return function $V(x(0), \alpha, u[0, N-1])$, about some nominal control sequence $u^j[0, N-1]$, and some nominal control parameter α^j, viz,

$$\begin{aligned} V(x(0), \alpha^{j+1}, u^{j+1}[0, N-1]) = {} & V(x(0), \alpha^j, u^j[0, N-1]) + V_\alpha(0)\,\delta\alpha \\ & + V_u(0)\,\delta u[0, N-1] + \tfrac{1}{2}\,\delta\alpha^{\mathrm{T}}\,V_{\alpha\alpha}(0)\,\delta\alpha \\ & + \delta\alpha^{\mathrm{T}}\,V_{\alpha u}(0)\,\delta u[0, N-1] \\ & + \tfrac{1}{2}\,\delta u^{\mathrm{T}}[0, N-1]\,V_{uu}(0)\,\delta u[0, N-1] \end{aligned} \tag{3.111}$$

where $\delta\alpha = \alpha^{j+1} - \alpha^j$ and $\delta u[0, N-1] = u^{j+1}[0, N-1] - u^j[0, N-1]$ must be small enough to ensure the validity of the expansion. In the successive sweep algorithm $\delta\alpha$ and δu are chosen to maximize the right-hand side, i.e.,

$$\begin{bmatrix} \delta u[0, N-1] \\ \delta\alpha \end{bmatrix} = -\begin{bmatrix} V_{uu} & V_{\alpha u} \\ V_{u\alpha} & V_{\alpha\alpha} \end{bmatrix}^{-1} \begin{bmatrix} V_u^{\mathrm{T}}(0) \\ V_\alpha^{\mathrm{T}}(0) \end{bmatrix} \tag{3.112}$$

One of the main drawbacks here is that such a choice would require the explicit inversion of a large matrix. Accordingly, as with the gradient method, a method is developed which chooses $\delta u(k)$ in a sequential manner.

First, we note that the Newton–Raphson choice for $\delta u^{j+1}[k, N-1]$ and hence for $\delta u(k)$ must maximize the second-order expansion of $V(x(k), \alpha^{j+1}, u^{j+1}[k, N-1])$, i.e.,

$$\begin{aligned} \delta u[k, N-1] = \arg \max_{\delta u[k, N-1]} \{ & V_x(k)\,\delta x(k) + V_\alpha(k)\,\delta\alpha \\ & + V_u(k)\,\delta u[k, N-1] + \tfrac{1}{2}\,\delta x(k)^{\mathrm{T}}\,V_{xx}(k)\,\delta x(k) \\ & + \delta x(k)^{\mathrm{T}}\,V_{xu}(k)\,\delta u[k, N-1] + \delta x(k)^{\mathrm{T}}\,V_{x\alpha}(k)\,\delta\alpha \\ & + \delta\alpha^{\mathrm{T}}\,V_{\alpha u}(k)\,\delta u[k, N-1] + \tfrac{1}{2}\,\delta\alpha^{\mathrm{T}} V_{\alpha\alpha}(k)\,\delta\alpha \\ & + \tfrac{1}{2}\,\delta u[k, N-1]^{\mathrm{T}}\,V_{uu}(k)\,\delta u[k, N-1]\} \end{aligned} \tag{3.113}$$

Here the expansion includes terms in $\delta x(k)$ to allow for changes in the state $x(k)$ due to changes in the previous controls $\delta u[0, k-1]$.

From this expansion it can be concluded that the Newton–Raphson choice of $\delta u[k, N-1]$, and hence of $\delta u(k)$, must be a linear function of $\delta x(k)$ and $\delta\alpha$, i.e.,

$$\delta u(k) = u_x(k)\,\delta x(k) + u_\alpha(k)\,\delta\alpha + \delta u_0(k) \tag{3.114}$$

This feedback law will be referred to as the *corrective feedback law*. In order to obtain the feedback coefficients for the kth stage, it is necessary to assume that the corrective feedback law has somehow been computed for the stages $k+1, \dots, N-1$. The return obtained by using a control $u(k)$ followed by the updated control $u^{j+1}(k+1)$ is given by $R(x(k), u(k), \alpha, k)$,

$$R(x(k), u(k), \alpha, k) = L(x(k), u(k), \alpha, k) + V^{j+1}(x(k+1), \alpha, k+1) \tag{3.115}$$

where $x(k+1) = F(x(k), u(k), \alpha, k)$. Now, expanding the return R to second order in terms of $\delta u(k)$, $\delta x(k)$, and $\delta\alpha$ about the nominal solution gives

$$R(x(k), \alpha^{j+1}, u^{j+1}(k)) = Z_0(k) + Z_x(k)\,\delta x(k) + Z_u(k)\,\delta u(k) + Z_\alpha(k)\,\delta\alpha$$
$$+ [\delta x^{\mathrm T}(k)\;\delta\alpha^{\mathrm T}\;\delta u^{\mathrm T}(k)]\begin{bmatrix} Z_{xx}(k) & Z_{x\alpha}(k) & Z_{xu}(k) \\ Z_{\alpha x}(k) & Z_{\alpha\alpha}(k) & Z_{\alpha u}(k) \\ Z_{ux}(k) & Z_{u\alpha}(k) & Z_{uu}(k) \end{bmatrix}\begin{bmatrix} \delta x(k) \\ \delta\alpha \\ \delta u(k) \end{bmatrix} \tag{3.116}$$

where

$$Z_0(k) = V^{j+1}(k+1) + L(k)$$
$$Z_x(k) = V_x^{j+1}(k+1)\,F_x(k+1) + L_x(k)$$
$$Z_u(k) = V_x^{j+1}(k+1)\,F_u(k+1) + L_u(k)$$
$$Z_\alpha(k) = V_x^{j+1}(k+1)\,F_\alpha(k+1) + L_\alpha(k) + V_\alpha(k+1)$$
$$Z_{xx}(k) = F_x^{\mathrm T}(k)\,V_{xx}^{j+1}(k+1)\,F_x(k) + V_x^{j+1}(k+1)\,F_{xx}(k) + L_{xx}(k)$$
$$Z_{xu}(k) = F_x^{\mathrm T}(k)\,V_{xx}^{j+1}(k+1)\,F_u(k) + V_x^{j+1}(k+1)\,F_{ux}(k) + L_{xu}(k)$$
$$Z_{uu}(k) = F_u^{\mathrm T}(k)\,V_{xx}^{j+1}(k+1)\,F_u(k) + V_x^{j+1}(k+1)\,F_{uu}(k) + L_{uu}(k)$$
$$Z_{x\alpha}(k) = F_x^{\mathrm T}(k)\,V_{xx}^{j+1}(k+1)\,F_\alpha(k) + F_x^{\mathrm T}(k)\,V_{x\alpha}^{j+1}(k+1)$$
$$+ V_x^{j+1}(k+1)\,F_{x\alpha}(k) + L_{x\alpha}(k)$$
$$Z_{u\alpha}(k) = F_u^{\mathrm T}(k)\,V_{xx}^{j+1}(k+1)\,F_\alpha(k) + F_u^{\mathrm T}(k)\,V_{x\alpha}^{j+1}(k+1)$$
$$+ V_x^{j+1}(k+1)\,F_{u\alpha}(k) + L_{u\alpha}(k)$$
$$Z_{\alpha\alpha}(k) = F_\alpha^{\mathrm T}(k)\,V_{xx}^{j+1}(k+1)\,F_\alpha(k) + F_\alpha^{\mathrm T}(k)\,V_{x\alpha}^{j+1}(k+1) + V_{\alpha x}^{j+1}(k+1)\,F_\alpha(k)$$
$$+ V_x^{j+1}(k+1)\,F_{\alpha\alpha}(k) + L_{\alpha\alpha}(k) \tag{3.117}$$

Now returning to the expansion, Equation (3.116), $\delta u(k)$ must be chosen to maximize the return $V(x(k), \alpha^{j+1}, u^{j+1}[k, N-1])$, i.e.,

$$\delta u(k) = -Z_{uu}^{-1}(k)[Z_{ux}(k)\,\delta x(k) + Z_{u\alpha}(k)\,\delta\alpha + Z_u(k)] \tag{3.118}$$

Next the sequential relations needed to form $V^{j+1}(k)$ and its partial derivatives are established. The control correction $\delta u(k)$,Equation (3.117), is substituted back into Equation (3.116) which gives

$$\begin{aligned}
V(x(k), \alpha^{j+1}, u^{j+1}(k, N-1)) = {} & V^{j+1}(k+1) - Z_u(k)\,Z_{uu}^{-1}(k)\,Z_u(k) \\
& + (Z_x(k) - Z_u(k)\,Z_{uu}^{-1}(k)\,Z_{ux}(k))\,\delta x(k) \\
& + (Z_\alpha(k) - Z_u(k)\,Z_{uu}^{-1}(k)\,Z_{u\alpha}(k))\,\delta\alpha \\
& + \tfrac{1}{2}\,\delta x^{\mathrm{T}}(k)(Z_{xx}(k) - Z_{xu}(k)\,Z_{uu}^{-1}(k)\,Z_{ux}(k))\,\delta x(k) \\
& + \delta x^{\mathrm{T}}(k)(Z_{x\alpha}(k) - Z_{xu}(k)\,Z_{uu}^{-1}(k)\,Z_{u\alpha}(k))\,\delta\alpha \\
& + \tfrac{1}{2}\,\delta\alpha^{\mathrm{T}}(Z_{\alpha\alpha}(k) - Z_{\alpha u}(k)\,Z_{uu}^{-1}(k)\,Z_{u\alpha}(k))\,\delta\alpha
\end{aligned} \tag{3.119}$$

Thus, taking a second-order expansion of the left-hand side about $x^j(k)$ and $\alpha^j(k)$ (u has just been chosen) and equating coefficients of identical terms in δx, $\delta\alpha$, etc., on both sides gives the following difference equations for the partial derivatives

$$\begin{aligned}
V^{j+1}(k) &= V^{j+1}(k+1) - Z_u(k)\,Z_{uu}^{-1}(k)\,Z_u(k) \\
V_x^{j+1}(k) &= Z_x(k) - Z_u(k)\,Z_{uu}^{-1}(k)\,Z_{ux}(k) \\
V_\alpha^{j+1}(k) &= Z_\alpha(k) - Z_u(k)\,Z_{uu}^{-1}(k)\,Z_{u\alpha}(k) \\
V_{xx}^{j+1}(k) &= Z_{xx}(k) - Z_{xu}(k)\,Z_{uu}^{-1}(k)\,Z_{ux}(k) \\
V_{x\alpha}^{j+1}(k) &= Z_{x\alpha}(k) - Z_{xu}(k)\,Z_{uu}^{-1}(k)\,Z_{u\alpha}(k) \\
V_{\alpha\alpha}^{j+1}(k) &= Z_{\alpha\alpha}(k) - Z_{\alpha u}(k)\,Z_{uu}^{-1}(k)\,Z_{u\alpha}(k)
\end{aligned} \tag{3.120}$$

These partial derivatives are computed as backward sequences starting at the final time where

$$\begin{gathered}
V_x(N) = \phi_x(N), \qquad V_\alpha(N) = \phi_\alpha(N) \\
V_{xx}(N) = \phi_{xx}(N), \qquad V_{x\alpha}(N) = \phi_{x\alpha}(N), \qquad V_{\alpha\alpha}(N) = \phi_{\alpha\alpha}(N)
\end{gathered} \tag{3.121}$$

Finally, at the initial stage, the change in the control parameter $\delta\alpha$ is obtained by maximizing the expansion of the return $V(x(0), \alpha, u^{j+1}(0, N-1))$ which gives

$$\delta\alpha = -V_{\alpha\alpha}^{-1}(0)[V_\alpha^{\mathrm{T}}(0) + V_{\alpha x}(0)\,\delta x(0)] \tag{3.122}$$

3.7.2 Summary of the Newton–Raphson Method

1. Choose a nominal control sequence $u^j[0, N-1]$ and control parameter α^j. Construct and store the trajectory $x[0, N]$ from the system equations (3.4) and the nominal control variables. Also compute the value of the performance index, Equation (3.5).

2. Compute the partial derivatives of V, V_x, $V_{x\alpha}$, V_{xx}, etc., from Equations (3.120) and (3.121).

3. Compute and store the feedback gains for the $\delta u[0, N]$ in Equation (3.118).

4. At the initial time compute α^{j+1} where $\alpha^{j+1} = \alpha^j + \delta\alpha$ and where $\delta\alpha$ is given in Equation (3.122).

5. Generate a new trajectory using the linear feedback law, i.e.,

$$u^{j+1}(k) = u^j(k) - Z_{uu}^{-1}(k)[Z_{ux}(k)(x^{j+1}(k) - x^j(k)) + Z_{u\alpha}(k)(\alpha^{j+1} - \alpha^j) + Z_u(k)]$$

and compute the cost. [If the cost is not reduced, a smaller step must be taken, e.g., replace $Z_u(k)$ by $\epsilon Z_u(k)$ where $0 < \epsilon < 1$.]

6. Return to step 2 and repeat until convergence is obtained.

If constraints of the form equation (3.6) are present, the performance index may be augmented, i.e.,

$$J^* = \phi(x(N), \alpha) + \nu^{\mathrm{T}}\psi(x(N), \alpha) + \sum_{i=0}^{N-1} [L(x(i), u(i), \alpha) + \nu^{\mathrm{T}}M(x(i), \alpha, u(i))]$$

where ν is the unknown Lagrange multiplier. Now, at the optimum for the constraints to be satisfied the return must be extremized with respect to ν. Hence, these multipliers may simply be considered as control parameters. Thus, the control parameters α are simply increased to include the Lagrange multipliers. Note, however, that an extremal is sought, not a maximum.

3.7.3 Examples

The Brachistochrone Problem. First, to compare the convergence of the successive sweep method with the gradient method, consider the brachistochrone problem that was described in detail earlier. The discrete system equation was $x(i+1) = x(i) + u(i)$ and the augmented performance index J^* is given by

$$J^* = \nu(x(N) + 0.5) + \sum_{i=0}^{N-1} h(i)[1 + u(i)^2]^{1/2}/[1 - x(i)]^{1/2}$$

Again, 100 equal steps $h(i)$ were chosen, i.e.,

$$h(i) = 0.01, \qquad i = 0, 1, \ldots, 99$$

The partial derivatives $L_x(i)$, $L_u(i)$, $F_x(i)$, and $F_u(i)$ are given by Equations (3.99) and the following additional partial derivatives are required,

$$V_x(N) = \nu, \qquad V_\nu(N) = x(N), \qquad V_{x\nu}(N) = 1$$

$$V_{xx}(N) = 0, \qquad V_{\nu\nu}(N) = 0$$

$$L_{xx}(i) = \tfrac{3}{4}h(i)(1 + u(i)^2)^{1/2}\,(1 - x(i))^{-5/2}$$

$$L_{xu}(i) = \tfrac{1}{2}h(i)\,u(i)(1 + u(i)^2)^{1/2}\,(1 - x(i))^{-3/2}$$

$$L_{uu}(i) = h(i)/(1 - x)^{1/2}\,(1 + u^2)^{3/2}$$

The nominal value for ν^j was taken to be 0. Thus, all the quantities needed to complete the algorithm are available.

The results of computations are portrayed in Tables X and XI. The optimal trajectory was identical to that obtained previously with the gradient algorithm (Figure 3.6). From Table X it can be seen that the fourth and fifth iterates yield good (six-place accuracy) approximations to the optimal solution, which was obtained by an independent calculation with the same program.

The following results, when compared with those obtained by gradient procedures in the previous section, demonstrate the advantages of the successive sweep. Far fewer iterations were required for convergence. In comparison with gradient methods, the successive sweep method provides rapid convergence. Theoretically, the number of accurate decimal places

TABLE X. BRACHISTOCHRONE PROBLEM

Iteration	Cost	$u(1)$	$x(101)$	ν
Nominal	1.1369745	0.0000000	−0.4999999	0.0000000
1	1.0012708	−0.6704293	−0.4999999	0.3510944
2	0.9995637	−0.7754882	−0.5000000	0.2254505
3	0.9995598	−0.7804138	−0.5000000	0.2229594
4	0.9995599	−0.7804242	−0.4999999	0.2229543
5	0.9995599	−0.7804242	−0.5000000	0.2229543

TABLE XI

BRACHISTROCHRONE PROBLEM; BACKWARD SWEEP QUANTITIES

t	u	x	Z_{uu}^{-1}	ν	V_{xx}	$V_{x\nu}$	$V_{\nu\nu}$
0	−0.780	0	0.495	0.624	0.468	0.947	−1.62
0.10	−0.707	−0.075	0.529	0.568	0.399	0.968	−1.44
0.20	−0.642	−0.142	0.561	0.515	0.330	0.989	−1.26
0.30	−0.584	−0.204	0.590	0.468	0.271	1.00	−1.09
0.40	−0.531	−0.260	0.616	0.426	0.219	1.01	−0.92
0.50	−0.482	−0.311	0.640	0.387	0.174	1.02	−0.75
0.60	−0.437	−0.357	0.662	0.351	0.133	1.02	−0.59
0.70	−0.395	−0.399	0.681	0.317	0.096	1.02	−0.44
0.80	−0.355	−0.436	0.699	0.285	0.062	1.02	−0.29
0.90	−0.317	−0.470	0.715	0.255	0.021	1.01	−0.14
1.0	−0.281	−0.500	0.728	0.233	0.000	1.00	0.00

should double with each successive iteration. In Table X the accurate digits are underlined. In general, the theoretical speed of convergence seems to be sustained.

An Orbit Transfer Problem. Let us consider a problem more closely related to modern technical problems, the orbit transfer problem. For the purpose of this problem, the rocket is constrained to a plane. The rocket position may be characterized by polar coordinates (r, ϕ), where r is the distance from the center of the gravitation field and ϕ is an angle (Figure 3.8). The velocity of the rocket may be expressed in terms of its components in the radial direction u and in the tangential direction v. The rocket will be assumed to have constant low thrust T. The only control that is available to the engineer is the direction of the thrust, which we measure by the angle θ that the rocket makes with the tangent to contours of constant radius. Figure 3.8 pictorally represents the variables just mentioned. At a particular point in time the dynamics of the rocket are governed by the following rules:

$$\dot{r} = u, \qquad \dot{u} = \frac{v}{r}$$

$$\dot{u} = \frac{v^2}{r} - \frac{\mu}{r^2} + \frac{T \sin\theta}{m_0 + \dot{m}(t - t_0)}, \qquad \dot{\phi} = \frac{uv}{r} + \frac{T\cos\theta}{m_0 + \dot{m}(t - t_0)}$$

The parameters appearing in the foregoing equations are μ equals MG, M equals the mass of the central body, G equals the gravitational constant, T is the thrust, m_0 is initial time, t_0 is initial time, and $-\dot{m}$ represents mass flow/unit time.

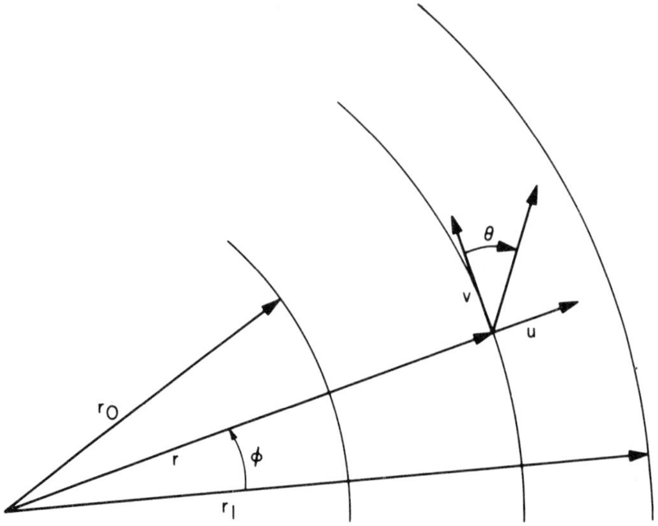

FIG. 3.8. The orbit transfer problem.

The problem that shall be considered here is the transfer of a spacecraft in a fixed time from an initial specified circular orbit to another circular orbit while maximizing the final radius.

To simplify the scaling the following normalized coordinates will be introduced

$$x_1 = \frac{r}{r_0}; \qquad x_2 = \frac{u}{(\mu/r_0)^{1/2}}; \qquad x_3 = \frac{v}{(\mu/r_0)^{1/2}}$$

$$t^1 = t\,\frac{(\mu)^{1/2}}{r_0^{\,3}}; \qquad u = \theta$$

In the new coordinates, the discrete form of the system equations is

$$x_1(i+1) = x_1(i) + h(i)\,x_2(i)$$

$$x_2(i+1) = x_2(i) + h(i)[x_3^{\,2}(i)/x_1(i) - 1/x_1^{\,2}(i) + A(i)\sin u(i)]$$

$$x_3(i+1) = x_3(i) + h(i)[-x_2(i)\,x_3(i)/x_1(i) + A(i)\cos u(i)]$$

where $A(i)$ equals $a/(1 - bt_i)$. To agree with Moyer and Pinkham (1964; see bibliography at end of Chapter 8) a and b are chosen as 1.405 and 0.07487, respectively. The initial boundary conditions become $x_1(0) = 1$, $x_2(0) = 0$, and $x_3(0) = 1$. The terminal boundary conditions are

$$\psi_{01} = x_2(N) = 0 \qquad \text{and} \qquad \psi_{02} = x_3(N) - (x_1(N))^{-1/2} = 0.$$

The final time $t_N = 3.32$ is chosen to agree with the earth—Mars minimal time according to the results obtained by Moyer and Pinkham (1964).

The modified performance index becomes

$$J^* = x_1(N) + \nu_1(x_2(N)) + \nu_2(x_3(N) - (x_1(N))^{-1/2})$$

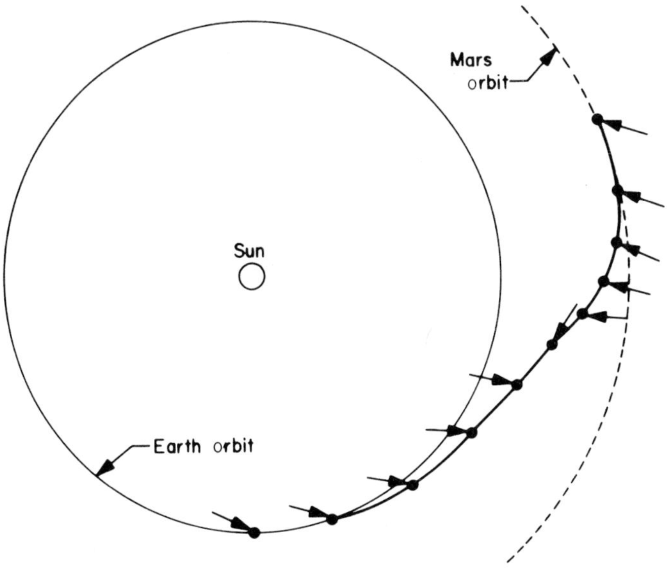

FIG. 3.9. Optimal transfer trajectory.

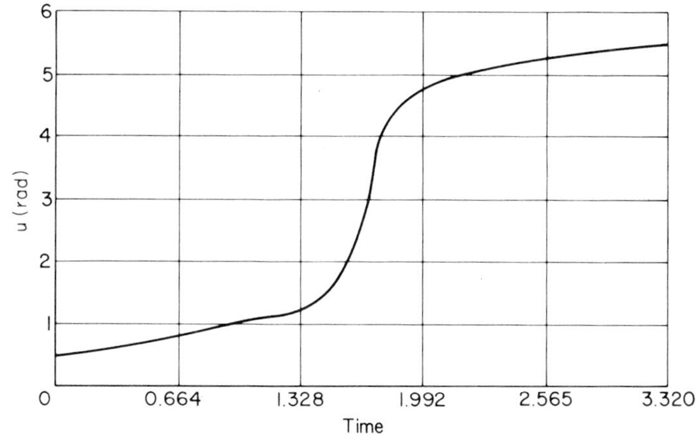

FIG. 3.10. Optimal control history.

The problem becomes that of maximizing J^* with respect to $u(i)$ $(i = 1, N)$ and extremizing J^* with respect to ν_1 and ν_2.

Numerical results were obtained using the discrete successive sweep algorithm.

One hundred equal time intervals were employed. Several nominals were used that yielded convergence to the optimal solution. The number of steps required to obtain three-place accuracy ranged from 13–18, depending upon the initial nominal. About 2 sec of IBM 7090 computer time was required per iteration.

The converged solution agrees with the results of others. Roughly three-place accuracy was obtained. The trajectory is graphed in Figure 3.9. The direction of the thrust is indicated at eleven equally spaced time intervals. The thrust program is graphed in Figure 3.10. Note that typical of low-thrust trajectories, the thrust has a positive radial component during the first half of the journey and then rapidly reverses itself to yield a negative radial component on the last half of the strip.

The optimal feedback gains along the optimal are tabulated in Table XII. These are the partial derivatives of the control with respect to the

TABLE XII. u, $x(1)$, $x(2)$, $x(3)$ ON THE OPTIMAL

t	u	$x(1)$	$x(2)$	$x(3)$
0	0.439	1.00	0.0	1.00
0.166	0.500	1.00	0.008	1.02
0.332	0.590	1.00	0.030	1.03
0.498	0.700	1.01	0.058	1.04
0.664	0.804	1.02	0.093	1.05
0.830	0.933	1.04	0.133	1.05
0.996	1.08	1.06	0.176	1.04
1.162	1.24	1.10	0.224	1.02
1.328	1.42	1.14	0.261	0.989
1.494	1.66	1.18	0.298	0.950
1.660	2.39	1.23	0.330	0.905
1.826	4.36	1.29	0.320	0.847
1.992	4.73	1.34	0.285	0.810
2.158	4.88	1.36	0.269	0.800
2.324	5.00	1.42	0.206	0.711
2.490	5.10	1.45	0.206	0.711
2.490	5.10	1.45	0.167	0.766
2.656	5.18	1.48	0.124	0.762
2.822	5.25	1.50	0.093	0.766
2.988	5.32	1.51	0.061	0.775
3.154	5.38	1.52	0.036	0.788
3.32	5.45	1.52	0.0	0.806

state and constraint levels ψ_1 ; ψ_2 . These derivatives are large on the first half of the trajectory, indicating that the optimal control function changes rapidly in this region.

3.7.4 The Convexity Condition

From a consideration of the second-order expansion of the return function equation (3.111), it is clear that the convexity condition is

$$\begin{bmatrix} \dfrac{\partial^2 V(x(0), \alpha, u[0, N-1])}{\partial u^2[0, N-1]} & \dfrac{\partial^2 V(x(0), \alpha, u[0, N-1])}{\partial \alpha\, \partial u[0, N-1]} \\ \dfrac{\partial^2 V(x(0), \alpha, u[0, N-1])}{\partial u[0, N-1]\, \partial \alpha} & \dfrac{\partial^2 V(x(0), \alpha, u[0, N-1])}{\partial \alpha^2} \end{bmatrix} \leqslant 0 \tag{3.123}$$

However, to test this condition would require an eigenvalue analysis of an $(mN + p) \times (mN + p)$ matrix. A somewhat simpler condition is suggested by the development of the successive sweep algorithm. It is clear from this development that for a maximum

$$Z_{uu}(k) \leqslant 0, \qquad V_{\alpha\alpha}(0) \leqslant 0, \qquad k = 0, 1, \ldots, N-1 \tag{3.124}$$

must hold and is equivalent to the condition just given. Here only an eigenvalue analysis of N $m \times m$ matrices plus one $p \times p$ matrix is required.

Classically the problem is viewed as a finite variable optimization problem with constraints. The convexity condition becomes

$$\delta^2 J^* \leqslant 0 \tag{3.125}$$

for all $\delta x(i)$, $\delta u(i)$, and $\delta\alpha$ where $\delta^2 J^*$ is the second-order term in an expansion of the cost J^*, Equation (3.105), i.e.,

$$\delta^2 J^* = \frac{1}{2}\left\{[\delta x^{\mathrm{T}}(N)\, \delta\alpha^{\mathrm{T}}] \begin{bmatrix} \phi_{xx} & \phi_{x\alpha} \\ \phi_{\alpha x} & \phi_{\alpha\alpha} \end{bmatrix} \begin{bmatrix} \delta x(N) \\ \delta\alpha \end{bmatrix} \right.$$
$$\left. + \sum_{t=0}^{N-1} [\delta x^{\mathrm{T}}(i)\, \delta u^{\mathrm{T}}(i)\, \delta\alpha^{\mathrm{T}}] \begin{bmatrix} H_{xx}(i) & H_{xu}(i) & H_{x\alpha}(i) \\ H_{ux}(i) & H_{uu}(i) & H_{u\alpha}(i) \\ H_{\alpha x}(i) & H_{ux}(i) & H_{\alpha\alpha}(i) \end{bmatrix} \begin{bmatrix} \delta x(i) \\ \delta u(i) \\ \delta\alpha \end{bmatrix} \right\} \tag{3.126}$$

and where the discrete Hamiltonian H was defined in Equation (3.104). The variations $\delta x(i)$, $\delta\alpha$, $\delta u(i)$ are governed by the equations

$$\begin{aligned} \delta x(i+1) &= F_x(i)\, \delta x(i) + F_u(i)\, \delta u(i) + F_\alpha(i)\, \delta\alpha \\ \delta x(0) &= 0 \end{aligned} \tag{3.127}$$

This formulation may be used to obtain the same convexity condition as in Equation (3.124). The approach is to seek a control parameter $\delta\alpha$ and a control sequence $\delta u[0, N-1]$ that maximizes $\delta^2 J^*$, Equation (3.126), subject to the system equations (3.127). This may be recognized as a linear quadratic problem to which the linear quadratic algorithm described earlier may be applied. The convexity condition for this problem [cf Equation (3.124)] is exactly equivalent to the convexity condition of Equation (3.125).

The strengthened convexity condition is that $Z_{uu}(k) < 0$ and $V_{\alpha\alpha}(0) < 0$ where k equals $0, 1, \ldots, N-1$ which implies that the second variation, $\delta^2 J$, is negative. This follows from the fact that the optimal solution to this problem [defined by Equations (3.126) and (3.127)] has the form

$$\delta\alpha^{\text{opt}} = -(V_{\alpha\alpha}^{\text{opt}}(0))^{-1}\, V_{\alpha x}^{\text{opt}}(0)\, \delta x(0)$$

$$\delta u^{\text{opt}}(i) = -Z_{uu}^{-1}(i)(Z_{ux}(i)\, \delta x(i) + Z_{u\alpha}(i)\, \delta\alpha)$$

Now, since $\delta x(0) = 0$, it may be shown easily that $\delta\alpha^{\text{opt}} = 0$, $\delta u^{\text{opt}}(i) = 0$ $(i = 0, \ldots, N-1)$, and $\delta x(i) = 0$ $(i = 0, \ldots, N)$. But, with these values for $\delta\alpha$, $\delta u(i)$, and $\delta x(i)$, $\delta^2 J^*$ is zero; i.e., the *upper bound* on the second variation is uniquely determined by the trivial perturbation $\delta\alpha = 0$, $\delta u(i) = 0$ $(i = 1, \ldots, N-1)$ and is equal to zero. Thus the second variation must be negative definite for all nontrivial perturbations.

Note that if $Z_{uu}(k)$ is singular, the sequential algorithm apparently fails because $Z_{uu}(k)^{-1}$ does not exist. However, the algorithm may still be employed by using the generalized inverse of $Z_{uu}(k)$, although in this case Z_{uu} is, at best, negative semidefinite.

3.7.5 The Discrete Maximum Principle: A Convexity Condition

The convexity condition obtained from the discrete maximum principle is $\partial^2 H(k)/\partial u^2(k) < 0$ where $k = 0, 1, \ldots, N-1$ or

$$p(k+1)\, F_{uu}(k) + L_{uu}(k) < 0$$

Comparing this condition with the convexity condition described in the previous section, namely, $Z_{uu}(k) < 0$ or

$$F_u^{\mathrm{T}}(k)\, V_{xx}(k)\, F_u(k) + V_x(k+1)\, F_{uu}(k) + L_{uu}(k) < 0$$

the reader can see that there is a difference between the two conditions; the term $F_u^{\mathrm{T}}(k)\, V_{xx}(k)\, F_u(k)$. Thus it is apparent that the discrete maximum principle will not necessarily be valid.

3.8 NEIGHBORING EXTREMAL METHODS

In many practical situations an approximation to the closed-loop solution is required even though this approximation may be valid only in the neighborhood of the optimal solution. For example, in the orbit transfer problem it may be more advantageous to correct deviations from the optimal trajectory with a feedback law than to recompute a new nominal trajectory at every stage using some numerical algorithm. One approach to this problem, which involves forming a *linear* approximation to the control law, is known as the *neighboring extremal* technique.

It is assumed that an optimal control sequence $u^{\text{opt}}[0, N-1]$ has been obtained by some means, e.g., with some numerical algorithm. If the solution is truly optimal, it may be further assumed that there exists an optimal control law $u^{\text{opt}}(x(k))$ such that

$$u^{\text{opt}}(k) = u^{\text{opt}}(x^{\text{opt}}(k)), \qquad k = 0, 1, \dots, N-1 \tag{3.128}$$

where $x^{\text{opt}}(k)$ is the optimal state at the kth stage. The approximation that is desired is

$$u^{\text{opt}}(x(k)) \simeq u^{\text{opt}}(k) + \frac{\partial u^{\text{opt}}}{\partial x(k)} (x^{\text{opt}}(k))\, \delta x(k) \tag{3.129}$$

where δx is small enough to ensure the validity of the expansion. The feedback law, for small perturbations about the optimal solution u^{opt}, x^{opt}, will then be given by $\delta u(\delta x(k)) = u_x\, \delta x(k)$.

The partial derivative u_x (an $n \times m$ matrix) is obtained from the dynamic programming equations. The return R was defined as $R = L(x(k), u(k)) + V^{\text{opt}}(x(k+1))$, where $x(k+1)$ is given by $x(k+1) = F(x(k), u(k))$. At the optimum, from Equation (3.25), R_u is zero, i.e.,

$$\partial R(k)/\partial u = V_x^{\text{opt}}(x(k+1))\, F_u(k) + L_u(k) = 0$$

Differentiation of this expression with respect to x gives

$$Z_{xu}(k) + Z_{uu}(k) \frac{\partial u^{\text{opt}}(k)}{\partial x(k)} = 0$$

where $Z_{xu}(k)$ and $Z_{uu}(k)$ are given by

$$\begin{aligned} Z_{xu}(k) &= F_x(k)\, V_{xx}^{\text{opt}}(k+1)\, F_u(k) + V_x^{\text{opt}}(k+1)\, F_{xu}(k) + L_{xu}(k) \\ Z_{uu}(k) &= F_u^{\text{T}}(k)\, V_{xx}^{\text{opt}}(k+1)\, F_u(k) + V_x^{\text{opt}}(k+1)\, F_{uu}(k) + L_{uu}(k) \end{aligned} \tag{3.130}$$

and where all the partial derivatives are evaluated at $x^{\text{opt}}(k)$ and $u^{\text{opt}}(k)$. Assuming Z_{uu} is not singular,

$$\frac{\partial u^{\text{opt}}(k)}{\partial x(k)} = -Z_{uu}^{-1}(k)\, Z_{ux}(k) \tag{3.131}$$

In order to compute u_x^{opt}, the partial derivatives V_x^{opt} and V_{xx}^{opt} must be computed. They are obtained by partially differentiating the dynamic programming equations (3.24) and (3.26) with respect to x, i.e.,

$$\begin{aligned} V_x^{\text{opt}}(N) &= \phi_x(N) \\ V_x^{\text{opt}}(k) &= L_x(x^{\text{opt}}(k), u^{\text{opt}}(k)) + V_x^{\text{opt}}(x^{\text{opt}}(k+1))\, F_x(k) \end{aligned} \tag{3.132}$$

and

$$\begin{aligned} V_{xx}^{\text{opt}}(N) &= \phi_{xx}(N) \\ V_{xx}^{\text{opt}}(k) &= Z_{xx}(k) + Z_{xu}(k)\, u_x(k) = Z_{xx}(k) - Z_{xu}(k)\, Z_{uu}^{-1}(k)\, Z_{ux}(k) \end{aligned} \tag{3.133}$$

where

$$Z_{xx}(k) = F_x^{\mathrm{T}}(k)\, V_{xx}^{\text{opt}}(k+1)\, F_x(k) + V_x^{\text{opt}}(k+1)\, F_{xx}(k) + L_{xx}(k) \tag{3.134}$$

Thus Equations (3.130)–(3.132), etc., allow the formation of the optimal feedback law, Equation (3.129).

Now returning to the equations derived in the previous section for the successive sweep method, we note a striking similarity. In fact, as the nominal solution approaches the optimal solution, the feedback gains used in the successive sweep algorithm become identical to those obtained before for the neighboring extremal algorithm. This similarity leads to the following observation: *The linear approximation to the optimal control law optimizes the second variation of the performance index.*

If terminal constraints are present the optimal feedback becomes

$$u^{\text{opt}}(x(k), \nu) = u^{\text{opt}}(k) + \frac{\partial u^{\text{opt}}(x(k), \nu)}{\partial x(k)}\, \delta x(k) + \frac{\partial u^{\text{opt}}(x(k), \nu)}{\partial \nu}\, \delta \nu \tag{3.135}$$

The approach taken here is to eliminate $\delta\nu$ in terms of δx. Note that

$$V_\nu = K \tag{3.136}$$

where K is the value of the constraint, i.e.,

$$\psi(x(N)) = K \tag{3.137}$$

Considering variations in Equation (3.136) gives $V_{\nu x}\, \delta x(k) + V_{\nu\nu}\, \delta\nu =$

δK. And assuming that the system is *not* abnormal, i.e., $V_{\nu\nu} \neq 0$, $\partial\nu$ is given by $\delta\nu = V_{\nu\nu}^{-1}[\delta K - V_{\nu x}\,\delta x(k)]$. Substituting this expression for $\delta\nu$ back into the expansion of u^{opt}, we obtain

$$u^{\text{opt}}(x(k), \nu) = u^{\text{opt}}(k) + [u_x^{\text{opt}}(k) - u_\nu^{\text{opt}}(k)\, V_{\nu\nu}^{-1}(k)\, V_{\nu x}(k)]\,\delta x(k) + u_\nu^{\text{opt}}(k)\, V_{\nu\nu}^{-1}\,\delta K$$

where the equations for $u_\nu^{\text{opt}}(k)$, $V_{\nu\nu}(k)$, etc. may be obtained from the dynamic programming equations.

Thus the neighboring optimal control law is linear in both the state and the level of the terminal constraint. In Table XIII the feedback gains are tabulated for the low-thrust example considered in the last chapter.

TABLE XIII. Feedback Gains on the Optimal

t	$u_{x(1)}$	$u_{x(2)}$	$u_{x(3)}$	u_{ψ_0}	u_{ψ_1}
0	1.62	0.937	1.33	−0.537	1.02
0.166	1.65	0.920	1.40	−0.604	1.06
0.332	1.70	0.935	1.52	−0.699	1.16
0.498	1.76	0.986	1.67	−0.828	1.28
0.664	1.83	1.08	1.89	−1.00	1.43
0.830	1.93	1.24	2.18	−1.29	1.61
0.996	2.06	1.49	2.62	−1.61	1.04
1.162	2.30	2.00	3.29	−2.21	2.16
1.328	2.84	2.92	4.89	−3.39	2.83
1.494	4.80	6.75	9.60	−7.39	5.88
1.660	21.88	52.9	44.2	−47.03	54.52
1.826	10.0	90.8	7.30	−9.40	75.00
1.992	−14.0	−76.9	−26.4	64.26	−80.30
2.158	−11.0	−83.4	−20.0	70.00	−86.70
2.324	−11.0	−83.4	−3.64	88.0	−103.1
2.490	−5.50	−104.0	15.0	111.0	−123.0
2.656	8.61	−183.0	44.6	162.7	−172.3
2.822	25.0	−280.0	112.0	250.0	−230.0
2.988	75.2	−548.0	292.0	510.6	−495.4
3.154	315.0	-1.42×10^3	$+1.2 \times 10^3$	1.8×10^3	-1.8×10^3
3.320	$+\infty$	$-\infty$	$+\infty$	$+\infty$	$-\infty$

PROBLEMS

1. The linear quadratic algorithm derived in Section 3.4.1 may be extended to include performance indices with linear terms. Show that if the

system equations are of the form $x(i+1) = F_x x(i) + F_u u(i)$ and the performance index has the form

$$J = x^{\mathrm{T}}(N)\,\phi_{xx}x(N) + \phi_x x(N) + \sum_{i=0}^{N-1} \{x^{\mathrm{T}}(i)\,L_{xx}(i)\,x(i)$$

$$+ x^{\mathrm{T}}(i)\,L_{xu}(i)\,u(i) + u^{\mathrm{T}}(i)\,L_{ux}(i)\,x(i) + u^{\mathrm{T}}(i)\,L_{uu}(i)\,u(i)$$

$$+ L_x(i)\,x(i) + L_u(i)\,u(i)\}$$

the optimal control law and the optimal return function have the following form

$$u^{\mathrm{opt}}(x(i)) = -D(i)\,x(i) - E(i)$$

$$V^{\mathrm{opt}}(x(i)) = x^{\mathrm{T}}(i)\,P(i)\,x(i) + Q(i)\,x(i) + R(i)$$

Derive sequential relations for $D(i)$ and $E(i)$.

2. Similarly, repeat the analysis of 3.4.5 using the generalized performance index.

3. Minimize the function J, where $J = \sum_{i=0}^{2} x(i)^2 + u(i)^2$ with $x(i+1) = x(i) + 0.1\,u(i)\; x(0) = 10$. (Here x and u are scalar variables.)

4. Repeat Problem 3 with the side constraint $x(3) = 0$.

5. Extend the analysis of 3.7.1 to include problems with terminal constraints. Show that the successive sweep correction to the nominal control becomes

$$\delta u(k) = -Z_{uu}^{-1}(k)[Z_{ux}(k)(x^{j+1}(k) - x^j(k)) + Z_{u\alpha}(k)(\alpha^{j+1} - \alpha^j) + Z_{u\nu}(k)(\nu^{j+1} - \nu^j) + Z_u(k)]$$

where ν is the multiplier used to adjoin the terminal constraints to the performance index.

6. Using the analysis from Problem 5 show that the return is *minimized* by the extremizing ν (for a maximization problem).

7. Find a simple example for which the discrete maximum principle does not hold.

8. Program the gradient and successive sweep algorithms and solve the brachistochrone problem.

BIBLIOGRAPHY AND COMMENTS

Section 3.3. The theory of dynamic programming was first announced formally in a paper by Bellman (1952):

Bellman, R. (1952). "The Theory of Dynamic Programming," *Proc. Natl. Acad. Sci. USA*, Vol. 38, pp. 716–719.

Subsequently several books have been written; for example, the reader should consult:

Bellman, R. (1957). *Dynamic Programming.* Princeton Univ. Press, Princeton, New Jersey.
Bellman, R. (1961). *Adaptive Control Processes: A Guided Tour.* Princeton Univ. Press, Princeton, New Jersey.
Bellman, R. and Dreyfus, S. (1962). *Applied Dynamic Programming.* Princeton Univ. Press, Princeton, New Jersey.

Section 3.4. The observation that the linear quadratic problem may be solved analytically has been exploited by a number of authors, for example,

Bellman, R. (1958). "Some New Techniques in the Dynamic Programming Solution of Variational Problems," *Quart. Appl. Math.*, Vol. 16, pp. 295–305.
Kalman, R. E. and Koepke, R. W. (1958). "Optimal Synthesis of Linear Sampling Control Systems Using Generalized Performance Indices," *Trans. A.S.M.E.*, Vol. 80, pp. 1820–1838.
Noton, A. R. M. (1965). *Introduction to Variational Methods in Control Engineering.* Pergamon, Oxford.
Tou, J. T. (1963). *Optimum Design of Digital Control Systems.* Academic Press, New York.
Wescott, J. H., ed. (1962). *An Exposition of Adaptive Control.* Pergamon, Oxford.

Section 3.5. Since the introduction of dynamic programming in 1952, a vast number of papers have appeared in the literature describing various computational techniques and applications. Apart from the books by Bellman (1957, 1961, and 1962), the reader may find the following references useful as a base for further study.

Aris, R. (1961). *The Optimal Design of Chemical Reactions: A Study in Dynamic Programming.* Academic Press, New York.
Aris, R. (1963). *Discrete Dynamic Programming.* Random House (Blaisdell), New York.
Bryson, A. E. and Ho, Y. C. (1969). *Optimization, Estimation and Control.* Random House (Blaisdell), New York.
Larson, R. E. (1967a). *State Increment Dynamic Programming.* American Elsevier, New York.
Larson, R. E. (1967b). "A Survey of Dynamic Programming Computational Procedures," *Trans. IEEE Autom. Control.*, Vol. AC-12, No. 6, pp. 767–774.
Roberts, S. M. (1964). *Dynamic Programming in Chemical Engineering and Process Control.* Academic Press, New York.

Section 3.6. Discrete gradient algorithms are discussed by

Bryson, A. E. and Ho, Y. C. (1969). *Optimization, Estimation and Control.* Random House (Blaisdell), New York.
Mayne, D. (1966). "A Second Order Gradient Method for Determining Optimal Trajectories of Non-Linear Discrete-Time Systems," *Intern. J. Control*, Vol. 3, pp. 85–95.

Wilde, D. J. and Beightler, C. S. (1967). *Foundations of Optimization.* Prentice-Hall, Englewood Cliffs, New Jersey.

A dynamic programming derivation is also given by Bellman and Dreyfus (1962).

Section 3.6.5. Early versions of the discrete maximum principle were given by

Chang, S. S. I. (1960). "Digitized Maximum Principle," *Proc. I.R.E.*, Vol. 48, pp. 2030–2031.

Katz, S. (1962). "A Discrete Version of Pontryagins Maximum Principle," *J. Elect. Control. (GB)*, Vol. 13, December.

and by:

Fan, L. T. and Wang, C. S. (1964). *The Discrete Maximum Principle.* Wiley, New York.

However, the assumptions made in the derivation of the principle were not made clear and some confusion resulted. As discussed in 3.6.5, the principle, in general, gives only a *stationarity* condition. Care must be taken when distinguishing between a maximum, a minimum, and a point of inflexion. These points were clarified in papers by the following:

Butkovskii, A. G. (1964). "The Necessary and Sufficient Conditions for Optimality of Discrete Control Systems," *Automation and Remote Control*, Vol. 24, pp. 963–970.

Halkin, H. (1964). "Optimal Control Systems Described by Difference Equations," *Advances in Control Systems: Theory and Applications* (C. T. Leondes, ed.). Academic Press, New York.

Halkin, H., Jordan, B. W., Polak, E., and Rosen, B. (1966). *Proc. 3rd I.F.A.C. Cong.* Butterworth, London and Washington, D.C. [See also the discussion in the book by Wilde and Beightler (1967) referenced before.]

Section 3.7. The successive sweep algorithm developed here is similar to ones obtained by

Bryson, A. and Ho, Y. C. (1969). *Optimization, Estimation and Control.* Random House (Blaisdell), New York.

McReynolds, S. R. (1967). "The Successive Sweep and Dynamic Programming," *J. Math. Anal. Appl.*, Vol. 19, No. 2.

Mayne, D. (1966). "A Second-Order Gradient Method for Determining Optimal Trajectories of Non-Linear Discrete Time Systems," *Intern. J. Control*, Vol. 3, pp. 85–95.

4

Optimization of Continuous Systems

In the remaining chapters the theory of dynamic programming is developed and applied to the optimization of continuous deterministic systems. Although this chapter is devoted to the derivation of the continuous dynamic programming equations and their application to linear quadratic problems, later chapters will consider the solution of nonlinear problems.

The dynamic programming approach will be developed formally via the finite difference technique. The technique of finite differences involves replacing the continuous problem by a sequence of multistage optimization problems and then considering these problems in the limit. This approach is preferred here because it quickly establishes the major results with a minimum amount of justification. Often, however, subsequent arguments or alternate proofs will be given to provide a more rigorous derivation.

As in the last chapter, a major emphasis is placed on the linear quadratic problem and several examples are given. One important example that is considered in some detail is the design of an aircraft autostabilizer.

4.1 NOTATION

4.1.1 System Variables

The system is assumed to be characterized by a finite set of real numbers that is referred to as the state of the system. The state at time t is denoted by an n-dimensional column vector $x(t)$, where

$$x(t) = \begin{bmatrix} x_1(t) \\ x_2(t) \\ \vdots \\ x_n(t) \end{bmatrix} \tag{4.1}$$

The control variables are classified in two sets. Variables of the one set are referred to as *control functions*, or, more simply, *controls*. These control functions, which are normally time varying, are denoted by any m-dimensional column vector of real numbers $u(t)$, where

$$u(t) = \begin{bmatrix} u_1(t) \\ u_2(t) \\ \vdots \\ u_m(t) \end{bmatrix} \tag{4.2}$$

The other set of control variables are *constant* and are referred to as *control parameters*. The control parameters are denoted by a p-dimensional vector α

$$\alpha = \begin{bmatrix} \alpha_1 \\ \alpha_2 \\ \vdots \\ \alpha_p \end{bmatrix} \tag{4.3}$$

4.1.2 System Dynamics

The dynamics of the system are expressed in terms of a set of first-order differential equations, viz,

$$\dot{x}(t) = f(x(t), u(t), \alpha)^{\dagger} \tag{4.4}$$

where $f^{\mathrm{T}} = (f_1, f_2, \ldots, f_n)^{\mathrm{T}}$ is an n-dimensional vector of functions.

4.1.3 Performance Index

The performance index is denoted by J

$$J = \phi(x(t_f), \alpha, t_f) + \int_{t_0}^{t_f} \{L(x(t), u(t), \alpha)\}\, dt \tag{4.5}$$

where L and ϕ are scalar, single-valued functions of their respective arguments. The final time t_f may be free or specified. In some cases,

† The overdot denotes differentiation with respect to time, for example, $dx/dt = \dot{x}$.

especially in stabilization problems, t_f may be regarded as being infinite. The initial time t_0 will usually be specified.

4.1.4 Side Constraints

The most general form of constraint is given by

$$0 = \psi(x(t_f), t_f) + \int_{t_0}^{t_f} \{M(x(t), u(t), \alpha)\} \, dt \tag{4.6}$$

where ψ and M may be scalar-valued or vector functions. If the function $M = 0$, the foregoing is referred to as a *terminal constraint*. Along with this *equality* constraint, inequality constraints may also have to be considered. In particular, constraints on the state of the form

$$S(x(t)) \leqslant 0 \tag{4.7}$$

and on the control variables

$$C(x(t), u(t)) \leqslant 0 \tag{4.8}$$

are very common. The functions C and S may be scalar-valued or vector functions. Constraints may also be placed on the initial state, although in the analysis that follows the initial state is assumed specified.

4.2 THE PROBLEM

In the light of these definitions, it is now possible to define the problem in a precise manner, viz: To find the control function $u(t)$ $t_0 \leqslant t \leqslant t_f$ and the control parameters α that maximize the performance index, Equation (4.5), subject to the state system equations (4.4) and the constraints, for example, Equations (4.6)–(4.8).

4.3 DYNAMIC PROGRAMMING SOLUTION

A natural approach is to replace the continuous problem by its finite difference approximation. The results of the previous chapter may then be applied. This technique of finite difference approximations was, in fact, used by Bernoulli and Euler in some of the early investigations into the calculus of variations.

The interval $[t_0, t_f]$ is divided into N equal intervals of time h. It is assumed that the control $u(i)$ is constant over the ith interval and the state $x(i)$ refers to the value of the state at the beginning of the ith interval or stage. The finite difference approximation to the system equation is then

$$x(i+1) = x(i) + hf(x(i), \alpha, u(i), t_i) \tag{4.9}$$

with $x(t_0)$ specified. The approximation to the performance index is

$$J = \varphi(x(t_f), \alpha, t_N) + \sum_{i=0}^{N-1} \{L(x(i), \alpha, u(i), t_i)\, h\} \tag{4.10}$$

Now following the development in the last chapter, the return $V(x(k), \alpha, u(k), t_k)$ is introduced, i.e.,

$$V(x(k), \alpha, u(k), t_k) = hL(x(k), \alpha, u(k), t_k) + V(x(k+1), \alpha, u(k+1), t_{k+1}) \tag{4.11}$$

where $x(k+1)$ is given by Equation (4.9). Note that the time $t_k = kh + t_0$ is included among the arguments of V. The optimal solution is given by u^{opt} and V^{opt} where

$$V^{\text{opt}}(x(N), \alpha, t_N) = \phi(x(N), \alpha, t_f) \tag{4.12}$$

$$V^{\text{opt}}(x(k), \alpha, t_k) = hL(x(k), \alpha, u^{\text{opt}}(k)) + V^{\text{opt}}(x(k+1), \alpha, t_{k+1}) \tag{4.13}$$

where $x(k+1) = x(k) + hf(x(k), \alpha, u^{\text{opt}}(k), t_k)$ and

$$u^{\text{opt}}(x(k), \alpha, t_k) = \arg\max_{u(k)} \{hL(x(k), \alpha, u(k), t_k) + V^{\text{opt}}(x(k+1), \alpha, t_{k+1})\} \tag{4.14}$$

where $x(k+1)$ is explicitly eliminated by

$$x(k+1) = x(k) + hf(x(k), \alpha, u(k), t_k)$$

The optimal value of the parameters α^{opt} is given by

$$\alpha^{\text{opt}} = \arg\max_{\alpha} V^{\text{opt}}(x(t_0), \alpha, t_0) \tag{4.15}$$

The continuous dynamic programming equations may now be obtained by letting $h \to 0$ and taking the limit of Equations (4.12)–(4.15). In the limit, the optimal control law and the optimal return junction for the discrete problem will approach the optimal control and return for the continuous problem. Equation (4.12) is unaltered and Equation (4.13) becomes a partial differential equation. First the return $V^{\text{opt}}(x(k+1), \alpha, t_{k+1})$ is expanded about $V^{\text{opt}}(x(k), \alpha, t_{k+1})$, viz,

$$V^{\text{opt}}(x(k+1), \alpha, t_{k+1}) = V^{\text{opt}}(x(k), \alpha, t_{k+1}) + V_x^{\text{opt}}(x(k), \alpha, t_{k+1}) \times f(x(k), \alpha, u^{\text{opt}}(k), t_k)h + O(h^2) \tag{4.16}$$

Substituting this expression into the right-hand side of Equation (4.13), we obtain

$$\begin{aligned} V^{\text{opt}}(x(k), \alpha, t_k) = {} & V^{\text{opt}}(x(k), \alpha, t_{k+1}) + hL(x(k), \alpha, u^{\text{opt}}(k)) \\ & + V_x^{\text{opt}}(x(k), \alpha, t_{k+1}) f(x(k), \alpha, u^{\text{opt}}(k), t_k)h \\ & + O(h^2) \end{aligned} \tag{4.17}$$

Now, dividing this equation by h gives

$$\begin{aligned} 0 = {} & [V^{\text{opt}}(x(k), \alpha, t_{k+1}) - V^{\text{opt}}(x(k), \alpha, t_k)]/h + L(x(k), \alpha, u(k), t_k) \\ & + V_x^{\text{opt}}(x(k), \alpha, t_{k+1}) f(x(k), \alpha, u^{\text{opt}}(k), t_{k+1}) + O(h) \end{aligned} \tag{4.18}$$

Hence, letting $h \to 0$, we obtain the following partial differential equation for V^{opt}

$$0 = V_t^{\text{opt}}(x, \alpha, t) + V_x^{\text{opt}}(x, \alpha, t) f(x, \alpha, u^{\text{opt}}, t) + L(x, u^{\text{opt}}, \alpha, t) \tag{4.19}$$

A similar analysis is employed to obtain $u^{\text{opt}}(x, \alpha, t)$. The expansion for $V(x(k+1), \alpha, t_{k+1})$, Equation (4.16), is substituted into Equation (4.14) and gives

$$\begin{aligned} u^{\text{opt}}(x(k), \alpha, t_k) = {} & \arg \max_{u(k)} \{V(x(k), \alpha, t_{k+1}) + h[L(x(k), \alpha, u(k), t_k) \\ & + V_x^{\text{opt}}(x(k), \alpha, t_{k+1}) f(x(k), \alpha, u(k), t_k)] + O(h^2)\} \end{aligned} \tag{4.20}$$

The first term does not depend upon $u(k)$, and as $h \to 0$ it is clear that

$$u^{\text{opt}}(x(k), \alpha, t_k) = \arg \max_{u(k)} \{L(x, \alpha, u, t) + V_x(x, \alpha, t) f(x, \alpha, u, t)\} \tag{4.21}$$

The dynamic programming solution may be summarized as follows.

1. Set

$$V^{\text{opt}}(x(t_f), \alpha, t_f) = \phi(x(t_f), \alpha, t_f) \tag{4.22}$$

2. Solve the partial differential equation

$$\begin{aligned} & V_t^{\text{opt}}(x(t), \alpha, t) + V_x^{\text{opt}}(x(t), \alpha, t) f(x(t), u^{\text{opt}}, \alpha, t) \\ & \qquad + L(x(t), \alpha, u^{\text{opt}}, t) = 0 \end{aligned} \tag{4.23}$$

where

$$u^{\text{opt}} = \arg \max_{u} \{V_x^{\text{opt}}(x, \alpha, t) f(x, u, \alpha, t) + L(x, u, \alpha, t)\} \tag{4.24}$$

3. At the initial time, solve for the control parameter α, i.e.,

$$\alpha^{\text{opt}}(x(t_0), t_0) = \arg\max_{\alpha}\{V^{\text{opt}}(x(t_0), \alpha, t_0)\} \tag{4.25}$$

4.3.1 Side Constraints

The dynamic programming equations for problems with side constraints may be developed by a similar limiting process. For example, consider terminal constraints of the form

$$0 = \psi(x(t_f), \alpha, t_f) \tag{4.26}$$

This is a special form of the more general constraint given in Equation (4.6). However, as was noted in 3.4.5, no generality is lost by considering this special case. As in the discrete problem, these constraints are adjoined to the performance index with a p-dimensional Lagrange multiplier ν. Thus the augmented performance index J^* is given by

$$J^* = \phi(x(t_f), \alpha, t_f) + \nu^{\mathrm{T}}\psi(x(t_f), \alpha, t_f) + \int_{t_0}^{t_f} L(x, u, \alpha, t)\, dt \tag{4.27}$$

Following an analysis similar to that just given for the unconstrained case, we obtain the following dynamic programming equations

$$V^{\text{opt}}(x, \alpha, \nu t_f) = \phi(x, \alpha, t_f) + \nu^{\mathrm{T}}\psi(x, \alpha, t_f) \tag{4.28}$$

$$0 = V_t^{\text{ext}} + V_x^{\text{ext}} f(x, u^{\text{ext}}, \alpha, t) + L(x, u^{\text{ext}}, \alpha, t) \tag{4.29}$$

$$u^{\text{ext}}(x, \alpha, \nu, t) = \arg\operatorname{ext}_u\{V_x^{\text{ext}} f(x, u, \alpha, t) + L(x, u^{\text{ext}}, \alpha, t)\} \tag{4.30}$$

$$\nu^{\text{opt}}(x(t_0), \alpha) = \arg\operatorname{ext}_\nu V^{\text{opt}}(x, \alpha, \nu, t_0) \tag{4.31}$$

$$\alpha^{\text{opt}}(x(t_0)) = \arg\max_\alpha V^{\text{opt}}(x, \alpha, \nu^{\text{opt}}(x, \nu), t_0) \tag{4.32}$$

Note that the *extremal* rather than optimal values for the control and return are given by these equations. The choice of ν ensures that the terminal constraint is satisfied. If a solution for ν does not exist or is not unique, the problem is said to be *abnormal*. Abnormality and its implications will be discussed later in this chapter, and it is assumed here that the problem is not abnormal.

Of course the equality

$$V_\nu^{\text{ext}}(x(t), \alpha, \nu, t) = 0 \tag{4.33}$$

must hold along the whole optimal trajectory. Hence, the multiplier ν may be eliminated at time $t > t_0$. If this is possible, the function

$\nu^{\text{ext}}(x, \alpha, t)$ may be obtained. Then the optimal control law and return function are obtained from the relations,

$$V^{\text{opt}}(x, \alpha, t) = V^{\text{ext}}(x, \alpha, \nu^{\text{ext}}(x, \alpha, t), t)$$

$$u^{\text{opt}}(x, \alpha, t) = u^{\text{ext}}(x, \alpha, \nu^{\text{ext}}(x, \alpha, t), t)$$

Note that in these circumstances V_x^{opt} is given by

$$V_x^{\text{opt}} = V_x^{\text{ext}} + V_\nu^{\text{ext}}\, \partial\nu/\partial x = V_x^{\text{ext}}$$

Hence the ν's may be eliminated and the dynamic programming equations become

$$0 = V_t^{\text{opt}}(x, \alpha, t) + L(x, u^{\text{opt}}, \alpha, t) + V_x^{\text{opt}}(x, \alpha, t) f(x, u^{\text{opt}}, \alpha, t) \tag{4.34}$$

and

$$u^{\text{opt}}(x, \alpha, t) = \arg\max_u \{L(x, u, \alpha, t) + V_x^{\text{opt}}(x, \alpha, t) f(x, u, \alpha, t)\} \tag{4.35}$$

4.3.2 An Alternative Derivation

The following derivation is somewhat more rigorous in its mathematical development, and, at the same time, illustrates the direct application of the optimality principle to continuous systems.

First, the partial differential equation for $V^{\text{opt}}(x, \alpha, t)$ will be obtained. Now V^{opt} is defined by

$$V^{\text{opt}}(x, \alpha, t) = \phi(x(t_f), \alpha, t_f) + \int_t^{t_f} L(x(s), u^{\text{opt}}(s), \alpha, s)\, ds \tag{4.36}$$

Hence,

$$V^{\text{opt}}(x, \alpha, t) = V^{\text{opt}}(x + \Delta x, \alpha, t + dt) + \int_t^{t+dt} L(x(s), u^{\text{opt}}(s), \alpha, s)\, ds \tag{4.37}$$

where, from the state equation (4.4)

$$\Delta x = \int_t^{t+dt} f(x(s), u(s), \alpha, s)\, ds \tag{4.38}$$

Now the term $V^{\text{opt}}(x + \Delta x, \alpha, t + dt)$ is expanded in a Taylor series to second order, i.e.,

$$V^{\text{opt}}(x, \alpha, t + dt) = V^{\text{opt}}(x, \alpha, t + dt) + V_x^{\text{opt}}(x, \alpha, t + dt)\, \Delta x + \Delta x^{\text{T}} V_{xx}^{\text{opt}}(\theta x + (1 - \theta)(x + \Delta x), \alpha, t + dt)\, \Delta x \tag{4.39}$$

where

$$0 \leqslant \theta \leqslant 1$$

and where the subscripts, as usual, denote partial derivatives. Under the assumption that $f(x(t), \alpha, u^{\text{opt}}(t), t)$ and V_{xx}^{opt} are bounded in the region of concern, the last term is of order $(dt)^2$. Thus, substituting this expansion into Equation (4.37) gives

$$V^{\text{opt}}(x, \alpha, t) - V^{\text{opt}}(x, \alpha, t + dt) = \int_t^{t+dt} L(x(s), u^{\text{opt}}(s), \alpha, s)\, ds + V_x^{\text{opt}}(x, \alpha, t + dt)\, \Delta x + 0(dt)^2 \quad (4.40)$$

Now dividing both sides of the foregoing equation by dt, taking the limit as $dt \to 0$, and noting that

$$\lim_{dt\to 0} \frac{\Delta x}{dt} = f(x(t), \alpha, u^{\text{opt}}(t), t)$$

and

$$\lim_{dt\to 0} \frac{1}{dt} \int_t^{t+dt} L(x(s), u^{\text{opt}}(s), \alpha, s)\, ds = L(x(t), \alpha, u^{\text{opt}}(t), t)$$

it is clear that V_t^{opt} exists and is given by

$$-V_t^{\text{opt}}(x, \alpha, t) = L(x(t), \alpha, u^{\text{opt}}(t), t) + V_x^{\text{opt}}(x, \alpha, t) f(x(t), u^{\text{opt}}(t), \alpha, t) \quad (4.41)$$

Note that in taking this limit the functions L, u, and V_x are required to be continuous. The boundary condition for $V^{\text{opt}}(x, \alpha, t)$ is provided by the identity

$$V^{\text{opt}}(x, \alpha, t_f) = \phi(x(t_f), \alpha, t_f) \quad (4.42)$$

Equations (4.41) and (4.42) are, of course, the same as the equations [(4.23) and (4.22)] just derived.

Next, the optimal control law is derived. Use is made of the obvious fact that any deviation from the optimal control *must* result in a lower value of the performance index. Suppose that the optimal control has been computed on the interval $t + dt \to t_f$, and that an arbitrary bounded continuous control $\tilde{u}$ is used on the interval $t \to t + dt$. From Equation (4.37),

$$V(x, \alpha, t, \tilde{u}) = \int_t^{t+dt} L(x(s), \alpha, \tilde{u}, s)\, ds + V^{\text{opt}}(x + \Delta x, \alpha, t + dt) \quad (4.43)$$

where

$$\Delta x = \int_t^{t+dt} f(x(s), \alpha, \tilde{u}(s), s)\, ds \quad (4.44)$$

Now as just stated, from the definition of a maximum,

$$V(x, \alpha, t, \tilde{u}) \leqslant V^{\text{opt}}(x, \alpha, t) \tag{4.45}$$

Hence,

$$-V^{\text{opt}}(x, \alpha, t) + V^{\text{opt}}(x + \Delta x, \alpha, t + dt) + \int_t^{t+dt} L(x(s), \alpha, \tilde{u}, s)\, ds \leqslant 0 \tag{4.46}$$

A Taylor series expansion of $V^{\text{opt}}(x + \Delta x, \alpha, t + dt)$ was given in Equation (4.39). Note, however, that here $f(x(t), u(t), \alpha, t)$ is replaced by $f(x(t), \tilde{u}, \alpha, t)$. Substituting this expansion into Equation (4.46) and dividing by dt gives

$$\{V^{\text{opt}}(x, \alpha, t + dt) - V^{\text{opt}}(x, \alpha, t)\}/dt + V_x^{\text{opt}}(x, \alpha, t + dt) f(x(t), \alpha, \tilde{u}, t)$$
$$+ \int_t^{t+dt} \frac{L(x(s), \alpha, \tilde{u}, s)\, ds}{dt} + 0(dt) \leqslant 0 \tag{4.47}$$

which in the limit as $dt \rightarrow 0$ becomes

$$V_t^{\text{opt}}(x, \alpha, t) + V_x^{\text{opt}}(x, \alpha, t) f(x(t), \alpha, \tilde{u}, t) + L(x(t), \alpha, \tilde{u}, t) \leqslant 0 \tag{4.48}$$

But $V_t^{\text{opt}}(x, \alpha, t)$ must satisfy Equation (4.41). Hence,

$$V_x^{\text{opt}}(x, \alpha, t) f(x(t), \alpha, \tilde{u}, t) + L(x(t), \alpha, \tilde{u}, t)$$
$$\leqslant V_x^{\text{opt}}(x, \alpha, t) f(x(t), \alpha, u^{\text{opt}}(t), t) + L(x(t), \alpha, u^{\text{opt}}, t) \tag{4.49}$$

In other words,

$$u^{\text{opt}}(x(t), \alpha, t) = \arg\max_u \{V_x^{\text{opt}}(x, \alpha, t) f(x(t), u(t), \alpha, t) + L(x(t), \alpha, u(t), t)\} \tag{4.50}$$

which is the same as Equation (4.24).

4.3.3 A Discussion of the Dynamic Programming Equations

The partial differential equation for the return function V^{opt}, Equation (4.23), is a generalization of the so-called Hamilton–Jacobi equation of classical mechanics. A similar equation may be derived for an arbitrary control law $u(x, t)$. For, if the return function $V(x, t)$ is defined by the equation

$$V(x_0, t_0) = \phi(x(t_f), t_f) + \int_{t_0}^{t_f} L(x(t), u(x, t), t)\, dt \tag{4.51}$$

where t_0 is the initial time and $x(t)$ is obtained by using the control $u(x, t)$, i.e., $\dot{x} = f(x, u(x, t), t)$ with the initial state $x(t_0) = x_0$, it is easy to show that

$$0 = V_t(x, t) + V_x(x, t) f(x, u(x, t), t) + L(x, u(x, t), t)$$
$$V(x(t_f), t_f) = \phi(x(t_f), t_f) \tag{4.52}$$

This equation is also referred to as the Hamilton–Jacobi partial differential equation.

Now the first two terms on the right-hand side of this equation correspond to the total derivative of V with respect to time, i.e.,

$$dV(x, t)/dt = V_t + V_x f \tag{4.53}$$

Hence, Equation (4.52) may be written

$$dV(x, t)/dt = -L(x, u(x, t), t) \tag{4.54}$$

Note that this result is a consequence of the definition of the return function equation (4.51). It is interesting to note that in the case where $L = 0$, the partial differential equation becomes

$$\dot{V} = \partial V^{\text{opt}}/\partial t + (\partial V^{\text{opt}}/\partial x) f = 0 \tag{4.55}$$

In other words, the optimal return function is a *constant* along the optimal trajectory.

Now suppose that at some point x^0 along an optimal trajectory, an arbitrary control $\tilde{u}$ is used in place of the optimal control. (See Figure 4.1 for a two-dimensional system.) Along the trajectory $\tilde{x}$ determined by $\tilde{u}$, the instantaneous rate of change of V^{opt} is given by

$$\partial V^{\text{opt}}/\partial t + (\partial V^{\text{opt}}/\partial x) f(x, \tilde{u}, t) \tag{4.56}$$

But from the dynamic programming equation (4.48) it follows that

$$\partial V^{\text{opt}}/\partial t + (\partial V^{\text{opt}}/\partial x) f(x, \tilde{u}, t) \leqslant 0 \tag{4.57}$$

Thus, we may conclude that the control $\tilde{u}$ cannot lead to an *increase* in the value of the return function. This property is clearly necessary for the optimal solution. It is also sufficient to guarantee the optimality of the solution as may be shown easily.

Referring to Figure 4.2, assume that over some arbitrary interval $[t_0, t_m]$, a *suboptimal* control $\tilde{u}(t)$ has been used in place of $u^{\text{opt}}(x, t)$. Let $\tilde{x}(t)$ denote the corresponding path. The return using the control $\tilde{u}$ along the path $x(t_0)$, $\tilde{x}(t_m)$ is $V^{\text{opt}}(\tilde{x}(t_m), t_m)$, whereas the return using the

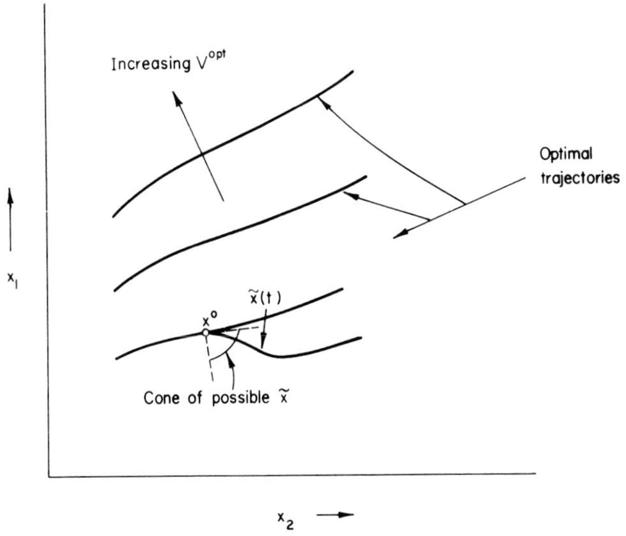

FIG. 4.1. The return associated with a nonoptimal trajectory $\tilde{x}$.

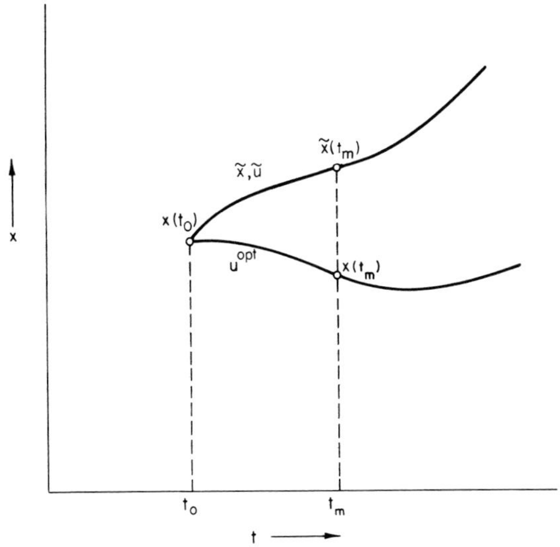

FIG. 4.2. Trajectory associated with the control $\tilde{u}$.

optimal control u^{opt} along the path $x(t_0) \to x(t_m)$ is $V^{\text{opt}}(x(t_0), t_0)$. Hence, the difference in return between the two paths is

$$\Delta J = V^{\text{opt}}(\tilde{x}(t_m), t_m) - V^{\text{opt}}(x(t_0), t_0)$$

Also the return using the control $\tilde{u}$ along the path $x(t_0)$, $\tilde{x}(t_m)$ must equal the optimal return $V^{\text{opt}}(x(t_0), t_0)$ added to the *integral of the rate of change of the optimal return along this path*, i.e.,

$$V^{\text{opt}}(\tilde{x}(t_m), t_m) = V^{\text{opt}}(x(t_0), t_0) + \int_{\tilde{x}} [V_t^{\text{opt}} + V_x^{\text{opt}} f] \, dt$$

and hence ΔJ is given by

$$\Delta J = \int_{\tilde{x}} [V_t^{\text{opt}} + V_x^{\text{opt}} f] \, dt \tag{4.58}$$

But, as just noted, the integrand is nonpositive and $\Delta J \leqslant 0$, which is the required result.

The extension of this argument to the case in which L is nonzero is left to the reader as an exercise.

It is interesting to note that in the previous proof no restrictions were imposed on the magnitude or duration of the perturbation. Thus, it follows that the existence of a global dynamic programming solution implies a *global optimum*. Unfortunately, in general such global solutions are often unattainable. However, tests for *local* existence may be established and they will be discussed in the following chapters.

4.3.4 Solution to the Dynamic Programming Equations

In general, as with the discrete dynamic programming equations, an analytic solution is not possible and recourse must be made to numerical techniques. These techniques will be discussed in later chapters. However, an analytic solution is possible for the linear quadratic problem. This solution, with examples, is the subject of the remainder of this chapter.

4.4 THE LINEAR QUADRATIC PROBLEM

The system equations are defined as

$$\dot{x}(t) = f_x x + f_u u, \qquad t_0 \leqslant t \leqslant t_f \tag{4.59}$$

The $n \times n$ matrix f_x and the $n \times m$ matrix f_u are system matrices which may be time varying. It is assumed that the initial conditions are specified, i.e.,

$$x(t_0) = x_0 \quad \text{specified} \tag{4.60}$$

The performance index is given by

$$J = \tfrac{1}{2}\, x^{\mathrm{T}}(t_f)\, \phi_{xx} x(t_f) + \tfrac{1}{2} \int_{t_0}^{t_f} \{x^{\mathrm{T}} L_{xx} x + u^{\mathrm{T}} L_{uu} u\}\, dt \tag{4.61}$$

where the symmetric weighting matrices L_{xx}, L_{uu}, ϕ_{xx} are dimensioned $n \times n$, $m \times m$, and $n \times n$, respectively.

Thus, the dynamic programming equations just derived become, for this problem,

$$V^{\mathrm{opt}}(x(t_f), t_f) = \tfrac{1}{2} x(t_f)\, \phi_{xx} x(t_f) \tag{4.62}$$

$$\frac{1}{2}\, x^{\mathrm{T}} L_{xx} x + \frac{1}{2}\, u^{\mathrm{opt}} L_{uu} u^{\mathrm{opt}} + \frac{\partial V^{\mathrm{opt}}}{\partial x} [f_x x + f_u u^{\mathrm{opt}}] + \frac{\partial V^{\mathrm{opt}}}{\partial t} = 0 \tag{4.63}$$

and

$$u^{\mathrm{opt}} = \arg \max_u \left\{ \frac{1}{2}\, x^{\mathrm{T}} L_{xx} x + \frac{1}{2}\, u^{\mathrm{T}} L_{uu} u + \frac{\partial V^{\mathrm{opt}}}{\partial x} (f_x x + f_u u) \right\} \tag{4.64}$$

where V^{opt} and u^{opt} are the optimal return and control, respectively.

Recalling the solution to the discrete problem, it is reasonable to assume that a solution exists for V^{opt} of the form

$$V^{\mathrm{opt}} = \tfrac{1}{2} x^{\mathrm{T}} V_{xx} x \tag{4.65}$$

where V_{xx} is an $n \times n$ matrix. Now Equation (4.64) can be rewritten by eliminating $\partial V^{\mathrm{opt}}/\partial x$, where

$$\partial V^{\mathrm{opt}}/\partial x = x^{\mathrm{T}} V_{xx} \tag{4.66}$$

giving

$$u^{\mathrm{opt}} = \arg \max_u \{\tfrac{1}{2} x^{\mathrm{T}} L_{xx} x + \tfrac{1}{2} u^{\mathrm{T}} L_{uu} u + x^{\mathrm{T}} V_{xx} (f_x x + f_u u)\} \tag{4.67}$$

Carrying out the indicated maximization gives

$$u^{\mathrm{opt}} = -L_{uu}^{-1} \{f_u{}^{\mathrm{T}} V_{xx} x\} \tag{4.68}$$

(It is tacitly assumed that the matrix L_{uu} is negative definite.) Now using Equations (4.65), (4.66), and (4.68) to eliminate $\partial V^{\mathrm{opt}}/\partial t$, $\partial V^{\mathrm{opt}}/\partial x$, and u^{opt} from Equation (4.63) gives

$$0 = \tfrac{1}{2} x^{\mathrm{T}} (L_{xx} + f_x{}^{\mathrm{T}} V_{xx} + V_{xx} f_x - V_{xx} f_u L_{uu}^{-1} f_u{}^{\mathrm{T}} V_{xx} + \dot{V}_{xx})\, x \tag{4.69}$$

Clearly, the original assumption, Equation (4.65), will be valid if $V_{xx}(t)$ is chosen so that

$$-\dot{V}_{xx} = L_{xx} + f_x{}^{\mathrm{T}} V_{xx} + V_{xx} f_x - V_{xx} f_u L_{uu}^{-1} f_u{}^{\mathrm{T}} V_{xx} \tag{4.70}$$

and

$$V_{xx}(t_f) = \phi_{xx} \tag{4.71}$$

The optimal control is, therefore, given by Equation (4.68) with V_{xx} given by Equations (4.70) and (4.71). Thus, as for the discrete linear quadratic problem, the optimal control consists of a linear combination of the state variables. From an engineering standpoint the result is very satisfying, as the implementation of the optimal control will be relatively straightforward (providing, of course, that all of the state variables can be measured).

It is true that Equation (4.70), the so-called matrix Riccati equation, must be integrated to obtain V_{xx} and also that an analytic solution for V_{xx} is rarely available. However, the equation may be integrated quite easily on a digital computer using, for example, a standard Runge–Kutta technique. Furthermore, V_{xx} will be symmetric and so only $n(n+1)/2$ equations have to be integrated.

4.4.1 Examples

The practical application of these results will now be illustrated with several problems. Although the first problem may be solved analytically, the other two are somewhat more complex and the matrix V_{xx} must be found by integrating Equation (4.70) on a digital computer. These examples involve the stabilization of a gyro accelerometer and the design of an aircraft autostabilizer.

EXAMPLE 4.1. *An Analytic Example.* The velocity of a low-thrust space probe, neglecting gravitational forces, is given simply by

$$\dot{x} = u \tag{4.72}$$

where x is the velocity and u is the thrust. A problem of interest is, given some initial value for x_0, to minimize the index J where

$$J = \tfrac{1}{2} \int_0^{t_f} \{x^2 + u^2\}\, dt \tag{4.73}$$

In terms of the notation just used,

$$f_x = 0, \quad f_u = 1, \quad L_{xx} = 1, \quad L_{uu} = 1, \quad \text{and} \quad \phi_{xx} = 0 \tag{4.74}$$

Hence, the Riccati equation is [cf Equation (4.70)],

$$\dot{V}_{xx} = V_{xx}^2 - 1, \qquad V_{xx}(t_f) = 0 \tag{4.75}$$

where here V_{xx} is a scalar. The solution is easily obtained. Let $\tau = t_f - t$ and Equation (4.75) becomes

$$dV_{xx}/d\tau = 1 - V_{xx}^2 \tag{4.76}$$

This equation may be integrated analytically to give

$$V_{xx}(\tau) = \tanh(\tau) \tag{4.77}$$

Hence, the optimal control [cf Equation (4.68)] is given by

$$u^{\text{opt}}(\tau) = -\tanh(\tau)\, x(\tau) \tag{4.78}$$

Note that as τ increases $\tanh(\tau)$ tends to a constant, i.e., $\lim \tau \to \infty$; $\tanh(\tau) = 1$. In other words, the optimal feedback gain is a constant for large values of τ, and, in practice a fixed feedback gain could be used over the whole interval $0 \to t_f$. Although this result is shown here for the scalar case, it may also be true for multivariable systems, at least when the matrices f_x, f_u, L_{xx}, and L_{uu} are constant.

In the case when $t_f = \infty$, the optimal feedback gain D is $D = \tanh(\infty) = 1$ for all finite time. Thus, $u^{\text{opt}}(x(t))$ is given by $u^{\text{opt}}(t) = -x(t)$. Substituting this value for u into the system equation gives $\dot{x}(t) = -x(t)$ and so $x(t) = x_0 e^{-t}$. Evaluating the performance index gives $J^{\text{opt}} = x_0^2 \int_0^\infty e^{-2t}\, dt = \frac{1}{2} x_0^2$ which confirms that the optimal cost is proportional to the initial state squared, i.e., $J^{\text{opt}} = \frac{1}{2} x_0^2 V_{xx}$.

EXAMPLE 4.2. *The Control of an Accelerometer.* The problem considered is that of stabilizing an accelerometer; in this example, one that belongs to the class of pendulous integrating gyro accelerometers. The action of the accelerometer may be described with reference to Figure 4.3 as follows. It consists of a gyrowheel, mounted inside a cylinder, rotating at an angular velocity ω_s, and establishing the gyroscopic angular momentum H. This first cylinder is supported inside a second concentric cylinder by means of an air bearing. The center of mass differs from the geometric support center in such a manner that if an acceleration is experienced along the input axis (IA), a torque T_β will be developed about the output axis (OA). This torque will be felt by the gyroscopic wheel and an angular motion about the input axis (IA) will be produced. If, due to friction or any other causes, an angular displacement β occurs between

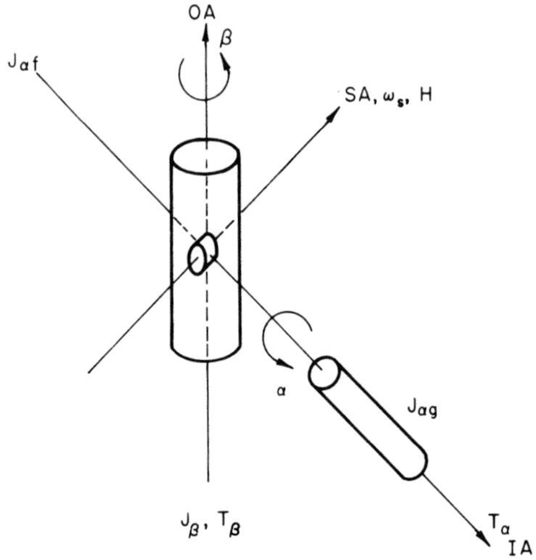

FIG. 4.3. Simplified line drawing of the K-8 accelerometer.

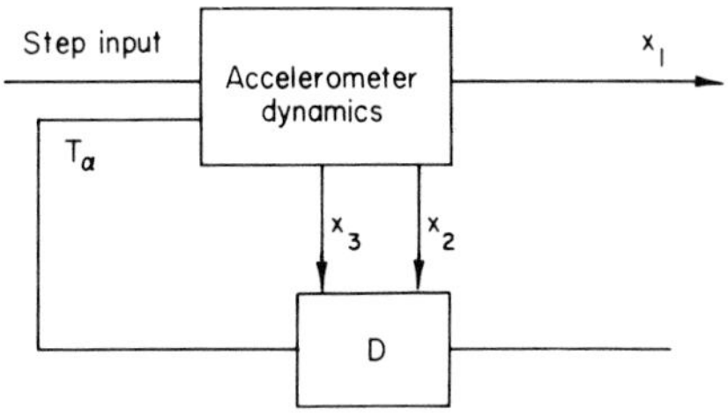

FIG. 4.4. Mechanization of the optimal controller.

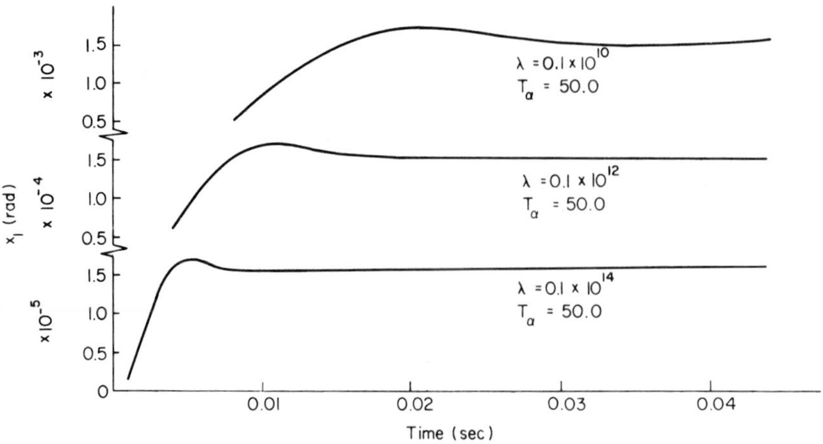

FIG. 4.5. The response of x_1 for a step input on T_α with λ as the parameter.

the two cylinders, a signal will be generated that will drive a torque motor located along the input axis (IA). This additional torque T_α will null out the angular displacement β. For a more detailed explanation of its construction and/or its operation, consult one of the numerous texts on gyroscopes [for example, those by Arnold and Maunder (1961), Fernandez and Macomber (1962)].

The problem is to minimize the displacement β given some arbitrary input torque T_α. Here, for the purpose of illustration, this torque will be taken to be a step function and it will be assumed that the acceleration torque T_β is zero.

The system equations are

$$\begin{aligned} \dot{x}_1 &= x_2 \\ \dot{x}_2 &= 756.097\ x_3 \\ \dot{x}_3 &= -34.96\ x_2 + 0.376\ u \end{aligned}$$

where, x_1 equals the angular displacement β (radius), x_2 equals the angular displacement rate $\dot{\beta}$ (radians per second), x_3 equals the angular displacement α (radians), and u equals the control torque T_α (grams-centimeter).

To ensure that the accelerometer obeys these equations, the magnitude of the displacement $\beta(x_1)$ should be less than 0.052 rad and the control torque T_α should be less than 1440 gm-cm.

Thus a suitable performance criterion is

$$J = \int_0^{t_f} (\lambda x_1^2 + u^2)\, dt$$

where the weighting parameter λ is chosen so the foregoing constraints are met; increasing λ will reduce the displacement x_1, although at the expense of increasing the control u. The final time t_f is chosen so that a steady-state solution to the Riccati equation is obtained. Here a suitable value was 0.01 sec.

The matrices needed to form the Riccati equation and hence the D matrix are

$$f_x = \begin{bmatrix} 0 & 1 & 0 \\ 0 & 0 & 756.097 \\ 0 & -34.96 & 0 \end{bmatrix}, \qquad f_u = \begin{bmatrix} 0 \\ 0 \\ 0.376 \end{bmatrix}$$

$$L_{xx} = \begin{bmatrix} \lambda & 0 & 0 \\ 0 & 0 & 0 \\ 0 & 0 & 0 \end{bmatrix} \qquad \text{and} \qquad L_{uu} = [1.0]$$

The optimal control was computed for various values of the weighting parameter λ. Then the controlled system was simulated on an analog computer and the response to a step input was examined. The mechanization of the controlled system is shown in Figure 4.4 and the responses for values of λ in the range $10^{10} < \lambda < 10^{14}$ are shown in Figure 4.5. The feedback gains are tabulated as a function of λ in Table XIV.

As can be seen, increasing λ quickens the response of the system while reducing the displacement x_1. All of these responses satisfy the constraints so that $\lambda \approx 10^{13}$ is probably the most suitable choice.

The Riccati equation was integrated backward using a Runge–Kutta integration scheme with a step length of 0.0001 sec, i.e., there were 100 steps. The computing time, including compilation and printout, was under 20 sec on the IBM 7040 computer.

TABLE XIV

TABULATION OF THE VALUES OF THE D-MATRIX AS A FUNCTION OF λ

λ	$-D(1, 1)$	$-D(1, 2)$	$-D(1, 3)$
0.1×10^{10}	−31,622.78	−194.45	−884.44
0.1×10^{11}	−100,000.00	−537.58	−1470.50
0.1×10^{12}	−316,227.78	−1290.69	−2278.53
0.1×10^{13}	−1,000,000.00	−2919.09	−3426.64
0.1×10^{14}	−3,162,277.80	−6429.97	−5085.68

EXAMPLE 4.3. *The Design of an Aircraft Autostabilizer.* Without some form of control system many dynamic systems are unstable or have undesirable properties. A typical example is a modern aircraft that may become extremely difficult to pilot without some form of control system.

Such a control system, usually referred to as an autostabilizer, must be suitable for use over the whole of the operating range of height and speed of the aircraft.

The dynamic equations of an aircraft are, of course, nonlinear, but a good approximation to the dynamics in level flight can be obtained by linearizing the equations at a given height and speed. In these circumstances the dynamics can be separated conveniently into sideways, or lateral motion and forward, or longitudinal motion. Here only the lateral mode will be considered as it presents the more interesting problem.

The linearized dynamics of the lateral mode of the aircraft are described by four first-order differential equations in roll rate $\dot{\phi}$, sideslip velocity V, yaw rate $\dot{\varphi}$, and roll angle ϕ [cf Figure 4.6]. The controls are the aileron

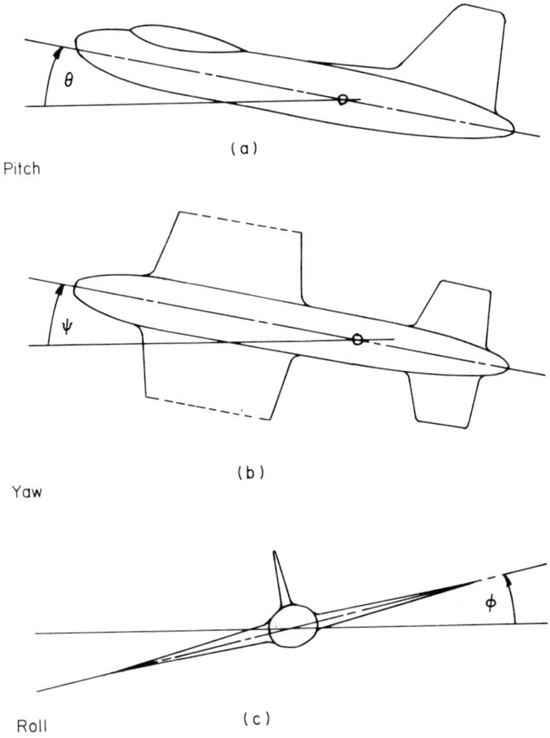

FIG. 4.6. Definition of angles. (a) is pitch, (b) is yaw, and (c) is roll.

angle ξ and the rudder angle ζ. Denoting the state variables by the vector $x = \{x_1, x_2, \ldots, x_6\}$ where

$$
\begin{aligned}
x_1 &= V, && \text{slideslip velocity (ft/sec)}\\
x_2 &= \phi, && \text{roll angle (rad)}\\
x_3 &= \dot{\phi}, && \text{roll rate (rad/sec)}\\
x_4 &= \dot{\psi}, && \text{yaw rate (rad/sec)}\\
x_5 &= \xi, && \text{aileron angle (rad)}\\
x_6 &= \zeta, && \text{rudder angle (rad)}
\end{aligned}
\tag{4.79}
$$

typical equations for a high speed transport at $M = 2$ and 60,000 ft are as follows.

$$\begin{aligned}
\dot{x}_1 &= -0.044x_1 + 38.1x_2 - 2240x_4 + 16.57x_6 \\
\dot{x}_2 &= x_3 \\
\dot{x}_3 &= -0.0027x_1 - 0.321x_3 + 1.55x_4 - 3.98x_5 + 0.996x_6 \\
\dot{x}_4 &= 0.0007x_1 - 0.122x_4 - 0.636x_5 - 0.636x_6 \\
\dot{x}_5 &= -10x_5 + 10u_1 \\
\dot{x}_6 &= -10x_6 + 10u_2
\end{aligned} \tag{4.80}$$

The fifth and sixth equations correspond to electrohydraulic actuators with an exponential time lag of 0.1 sec. u_1 and u_2 are the (electrical) command signals to the actuators.

One approach to the design of the autostabilizer is to ensure that the response of the controlled aircraft to a step or impulse command signal approximates that of some ideal or reference system. The ideal response can be considered to be that corresponding to a second-order system. Thus, denoting the desired roll rate by x_8 and its derivative by x_7, the desired system equations are

$$\dot{x}_7 = -4.2x_7 - 9x_8, \qquad \dot{x}_8 = x_7 \qquad (4.81)^\dagger$$

In addition to achieving a given response, it is assumed that (a) the normalized sideslip velocity, x_1/V_F,‡ must be minimized and (b) the input to the control actuators u_1 and u_2 must be constrained so that one degree change in the control angle produces about 1 deg/sec roll rate. A suitable performance index J may now be formulated, viz,

$$J = \int_0^T \{(x_8 - x_3)^2 + ax_1^2/V_F^2 + \lambda_1 u_1^2 + \lambda_2 u_2^2\}\, dt \tag{4.82}$$

The first term in the index places emphasis on the actual roll rate corresponding to the desired or model roll rate. The second term penalizes the sideslip V, and the third and fourth terms constrain the actuator signals. The performance index has been set up in terms of impulse response, as opposed to, say, the step response, because (a) a sustained command for roll rate is unrealistic and (b) use of the step response might place undue emphasis on matching the *steady-state* values of $\dot{\phi}$ and $\dot{\phi}_D$, (x_3 and x_8) which were not considered very important.

Numerical values now have to be assigned to the parameters T, a, λ_1, and λ_2. The integration time T is chosen so that the elements of the

† These equations correspond to a second-order system with an undamped natural frequency of 3 rad/sec and a damping ratio of 0.7.

‡ V_F is the forward speed.

D matrix reach a steady state, i.e., $T = \infty$; in this case $T = 5.0$ sec is used. The choice of values for the remaining constants presents little difficulty. The parameters may be determined quite logically; for example, increasing the parameter a will have the effect of reducing sideslip, but it will be at the expense of the other terms in the performance criterion; either larger control surface deflections will be required or the response of the aircraft will suffer. However, it was found that the choice of numerical values for the weighting parameters was *not* critical.

The following procedure was established to choose the correct weighting values.

1. The D matrix was computed with the parameters a, λ_1, and λ_2 set to nominal values.

2. The stabilized aircraft was simulated on an analog computer, and the response to an impulsive demand for roll rate was observed. The mechanization of the controller is shown in Figure 4.7.

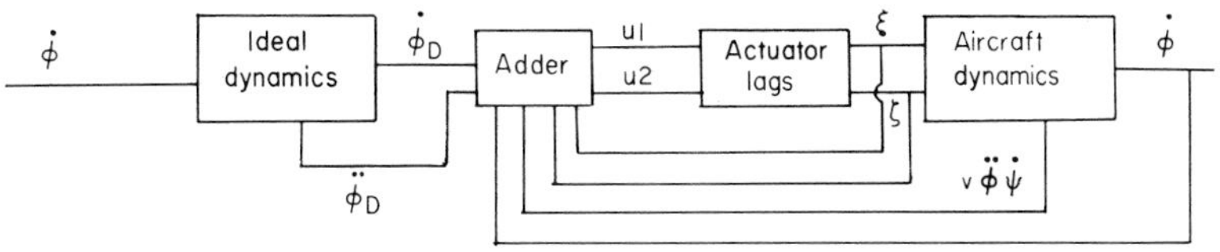

FIG. 4.7. Control system configuration for the lateral mode.

3. The values of the parameters a, λ_1, and λ_2 were adjusted as necessary, and the D matrix was recomputed.

The weighting on sideslip a was adjusted to minimize sideslip while maintaining correspondence between the actual and the ideal response, and λ_1 and λ_2 were chosen so that the forcing function amplitudes were within the specifications just outlined.

In this case, after a few computer runs a, λ_1, and λ_2 were chosen to be 1, 0.05, and 0.02, respectively, and with these values the matrices used in the computation of the D matrix were

$$L_{xx} = \begin{bmatrix} 1/V_F^2 & 0 & 0 & 0 & 0 & 0 & 0 & 0 \\ 0 & 0 & 0 & 0 & 0 & 0 & 0 & 0 \\ 0 & 0 & 1 & 0 & 0 & 0 & 0 & -1 \\ 0 & 0 & 0 & 0 & 0 & 0 & 0 & 0 \\ 0 & 0 & 0 & 0 & 0 & 0 & 0 & 0 \\ 0 & 0 & 0 & 0 & 0 & 0 & 0 & 0 \\ 0 & 0 & 0 & 0 & 0 & 0 & 0 & 0 \\ 0 & 0 & -1 & 0 & 0 & 0 & 0 & 1 \end{bmatrix}, \quad L_{uu} = \begin{bmatrix} 0.05 & 0 \\ 0 & 0.02 \end{bmatrix} \tag{4.83}$$

$$f_x = \begin{bmatrix} -0.044 & 38.0 & 0 & -2240 & 0 & 16.57 & 0 & 0 \\ 0 & 0 & 1 & 0 & 0 & 0 & 0 & 0 \\ -0.0027 & 0 & -0.321 & 1.55 & -3.98 & -996 & 0 & 0 \\ 0.0007 & 0 & 0 & 0.122 & -0.636 & -0.636 & 0 & 0 \\ 0 & 0 & 0 & 0 & -10 & 0 & 0 & 0 \\ 0 & 0 & 0 & 0 & 0 & -10 & 0 & 0 \\ 0 & 0 & 0 & 0 & 0 & 0 & -4.2 & -9 \\ 0 & 0 & 0 & 0 & 0 & 0 & 1 & 0 \end{bmatrix} \tag{4.84}$$

$$f_u = \begin{bmatrix} 0 & 0 \\ 0 & 0 \\ 0 & 0 \\ 0 & 0 \\ 10 & 0 \\ 0 & 10 \\ 0 & 0 \\ 0 & 0 \end{bmatrix} \tag{4.85}$$

and the optimal controller was found to be

$$\begin{aligned} u_1 &= 1.706x_1 - 0.036x_2 + 1.087x_3 + 3.37x_4 \\ &\quad -0.228x_5 - 0.115x_7 - 1407x_8 \\ u_2 &= 1.9x_1 - 0.020x_2 - 0.255x_3 + 1.4x_4 \\ &\quad + 0.080x_5 - 0.106x_6 + 0.06x_7 + 0.404x_8 \end{aligned} \tag{4.86}$$

The response of the aircraft with the control equations (4.86) is shown in Figure 4.8. This response clearly meets the design requirements just outlined.

4.4.2 Design Studies

The linearized state equations used in the foregoing computation are related to one particular set of aircraft operating conditions (speed equal to Mach 2 and height is 60,000 ft). It is clearly important to examine the optimum controller for a variety of speeds and altitudes. In this study, five representative cases were chosen. The numbers relate to a high-speed transport (cf the Concorde) designed to operate at 60,000 ft and Mach 2. The resultant controllers are shown in Table XV, and the impulse responses in Figures 4.9–4.11. It can be seen that there is very little

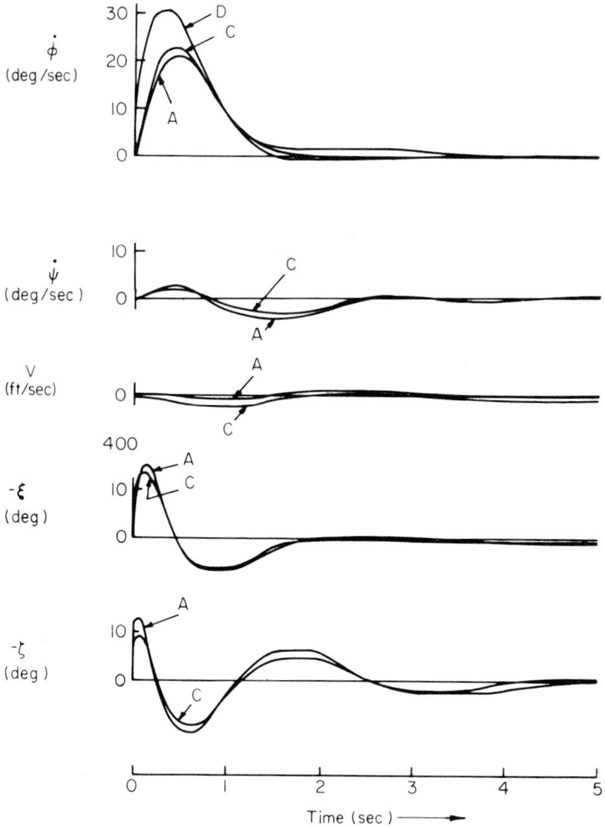

FIG. 4.8. High-speed transport responses. Case: 60,000 ft, $M = 2.0$. D is the ideal response, C the calculated controller, and A the average controller.

variation in the feedback coefficients in the different cases. This observation suggests that an average controller might be chosen, e.g.,

$$\begin{aligned} u_1 &= -1.99x_1 - 0.02x_2 + 1.10x_3 + 1.48x_4 - 0.79x_5 \\ &\quad + 0.07x_6 - 0.13x_7 - 1.00x_8 \\ u_2 &= -1.62x_1 - 0.28x_2 - 0.65x_3 + 2.86x_4 + 0.18x_5 \\ &\quad - 0.31x_6 + 0.07x_7 + 0.7x_8 \end{aligned} \tag{4.87}$$

The impulse responses of the high-speed transport with optimal and average controllers are also shown in Figures 4.8–4.11, and the average controller is indeed acceptable.

TABLE XV. COMPUTED CONTROLLERS FOR THE HIGH-SPEED TRANSPORT

	Aileron Feedback							
Case	V	$\dot{\phi}$	ϕ	$\dot{\psi}$	ξ	ζ	ϕ_D	$\dot{\phi}_D$
$M = 0.3$ Sea level	−1.860	1.125	−0.036	1.911	−0.302	0.004	−0.161	−0.971
$M = 0.5$ Sea level	−1.801	1.176	−0.020	1.477	−0.748	0.024	−0.138	−1.228
$M = 1.0$ Sea level	−2.409	1.197	−0.006	0.683	−2.784	0.288	−0.063	−1.331
$M = 1.0$ 60,000 ft	−1.526	1.288	−0.022	0.966	−0.429	0.035	−0.154	−1.044
$M = 1.5$ 60,000 ft	−1.347	0.697	−0.014	0.930	−0.283	0.008	−0.144	−0.957
$M = 2.0$ 60,000 ft	−3.008	1.119	−0.030	2.915	−0.203	0.051	−0.132	−0.506
	Rudder Feedback							
$M = 0.3$ Sea level	−2.723	−0.337	−0.319	4.606	0.020	−0.213	0.029	0.576
$M = 0.5$ Sea level	−1.626	−0.620	−0.131	2.128	0.122	−0.352	0.014	0.720
$M = 1.0$ Sea level	−0.234	−0.875	−0.025	0.562	1.438	−0.664	0.030	0.945
$M = 1.0$ 60,000 ft	−1.779	−0.889	−0.114	3.442	0.173	−0.175	0.117	0.833
$M = 1.5$ 60,000 ft	−1.072	−0.645	−0.048	2.158	0.040	−0.167	0.089	0.887
$M = 2.0$ 60,000 ft	−2.315	−0.521	−0.058	4.383	0.257	−0.285	0.111	1.216

As the flight cases for the aircraft were chosen at the extremes of the flight envelopes, it is reasonable to expect that the average controller, although suboptimal, would give satisfactory control over the whole of the flight envelope. It must be emphasized that involved techniques were *not* used to choose the average controller; it is quite possible that other suboptimal controllers could be chosen. In practice, engineering judgment and experience would help in the choice of the feedback elements from within the ranges indicated by the optimal controllers.

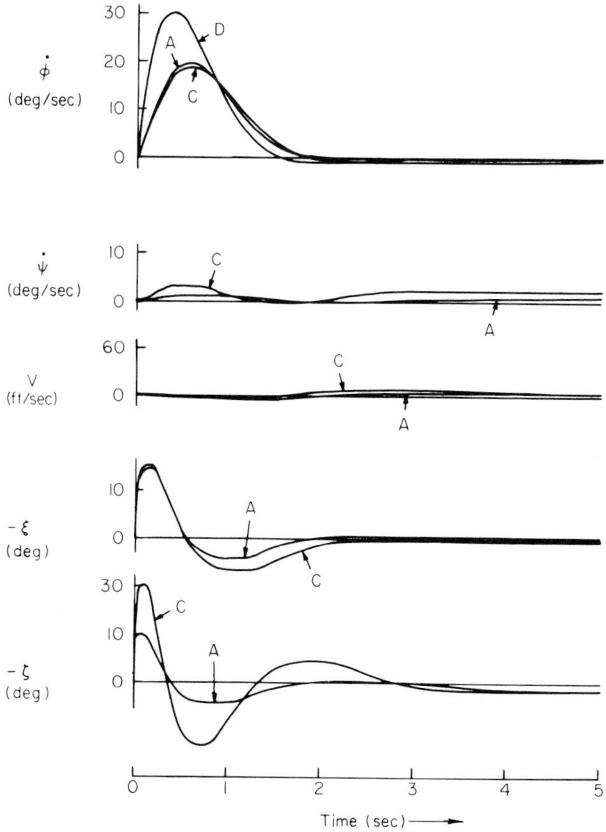

FIG. 4.9. High-speed transport responses. Case: sea level, $M = 0.3$. D is the ideal response, C the calculated controller, and A the average controller.

4.5 LINEAR QUADRATIC PROBLEM WITH CONSTRAINTS

One interesting variation of the linear quadratic problem that also admits a straightforward solution is the case where linear terminal constraints of the form

$$Mx(t_f) = d \tag{4.88}$$

are present. (M is a $p \times n$ matrix and d is a p-dimensional vector.) Now by adjoining these constraints to the performance index J, an augmented index J^* is formed where

$$J^* = \nu^{\mathrm{T}}(Mx(t_f) - d) + J \tag{4.89}$$

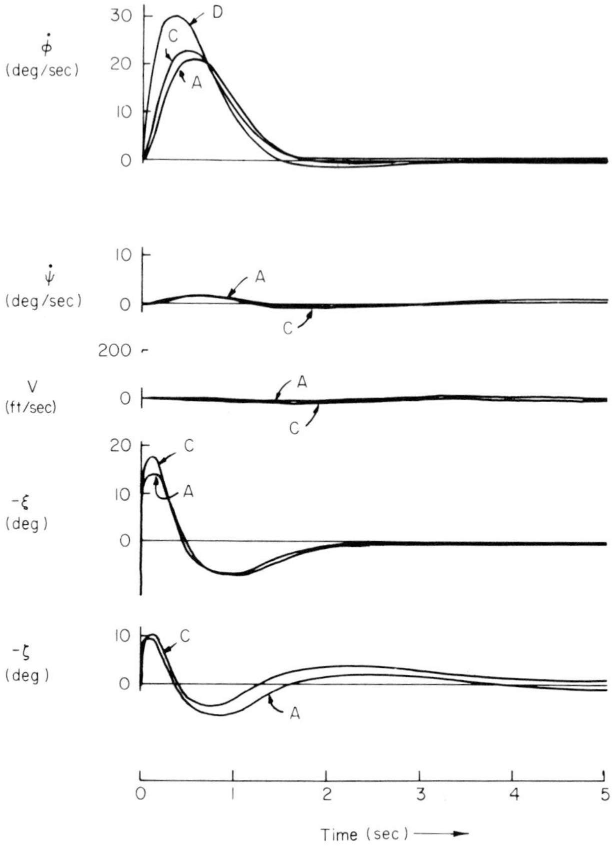

FIG. 4.10. High-speed transport responses. Case: 60,000 ft, $M = 1.0$. D is the ideal response, C the calculated controller, and A the average controller.

where ν is a p-dimensional vector of Lagrange multipliers. The dynamic programming equations are essentially the same as for the unconstrained problem

$$V^{\text{ext}}(x(t_f), t_f) = x^{\text{T}}\phi_{xx}x + \nu^{\text{T}}(Mx(t_f) - d) \tag{4.90}$$

$$\frac{1}{2}x^{\text{T}}L_{xx}x + \frac{1}{2}u^{\text{ext}}L_{uu}u^{\text{ext}} + \frac{\partial V^{\text{ext}}}{\partial x}(f_x x + f_u u^{\text{ext}}) + \frac{\partial V^{\text{ext}}}{\partial t} = 0 \tag{4.91}$$

and

$$u^{\text{ext}} = \arg\max_u \left\{\frac{1}{2}u^{\text{T}}L_{uu}u + \frac{\partial V^{\text{ext}}}{\partial x}(f_x x + f_u u)\right\} \tag{4.92}$$

Let us assume again that $V^{\text{ext}}(x, \nu, t)$ may be represented by a quadratic

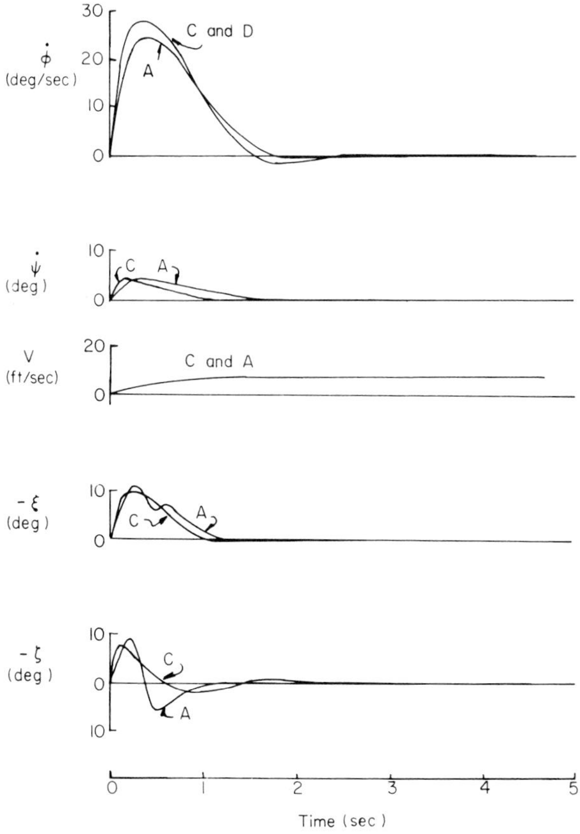

FIG. 4.11. High-speed transport responses. Case: sea level, $M = 1.0$. D is the ideal response, C the calculated controller, and A the average controller.

expansion except that now both linear and quadratic terms in x and ν are also included i.e.,

$$V^{\text{ext}} = V^0 + V_x x + V_\nu \nu + \frac{1}{2}[x^{\text{T}}, \nu^{\text{T}}] \begin{bmatrix} V_{xx} & V_{x\nu} \\ V_{\nu x} & V_{\nu\nu} \end{bmatrix} \begin{bmatrix} x \\ \nu \end{bmatrix} \tag{4.93}$$

where V_ν is a p-dimensional vector and $V_{x\nu}$ and $V_{\nu\nu}$ are $n \times p$ and $p \times p$ matrices. Hence, $\partial V^{\text{ext}}/\partial x$ is given by

$$\partial V^{\text{ext}}/\partial x = V_x + V_{xx} x + V_{x\nu} \nu \tag{4.94}$$

Substituting this expression into Equation (4.92) gives

$$u^{\text{ext}} = \arg\max\{\tfrac{1}{2}u^{\text{T}} L_{uu} u + (V_x{}^{\text{T}} + V_{xx} x + V_{x\nu} \nu)^{\text{T}} (f_x x + f_u u)\} \tag{4.95}$$

Now carrying out the indicated maximization by differentiating the right-hand side, setting to zero, and solving for u^{ext} gives

$$u^{\text{ext}} = -L_{uu}^{-1} f_u^{\mathrm{T}}(V_{xx}x + V_{x\nu}\nu + V_x) \tag{4.96}$$

Exactly as in the unconstrained case, u^{ext}, $\partial V^{\text{ext}}/\partial x$, and $\partial V^{\text{ext}}/\partial t$ may be substituted into the dynamic programming equation (4.91) which, after grouping related terms, becomes

$$\begin{aligned}
&\tfrac{1}{2}x^{\mathrm{T}}\{\dot{V}_{xx} + L_{xx} + f_x^{\mathrm{T}}V_{xx} + V_{xx}f_x - V_{xx}f_uL_{uu}^{-1}f_u^{\mathrm{T}}V_{xx}\}\,x \\
&\qquad + \{\dot{V}_x + f_x^{\mathrm{T}}V_x - V_{xx}f_uL_{uu}^{-1}f_u^{\mathrm{T}}V_x\}\,x + \{\dot{V}_\nu + V_xf_uL_{uu}^{-1}f_uV_{x\nu}\}\,\nu \\
&\qquad + x^{\mathrm{T}}\{\dot{V}_{x\nu} + f_x^{\mathrm{T}}V_{x\nu} - V_{xx}f_uL_{uu}^{-1}f_u^{\mathrm{T}}V_{x\nu}\}\,\nu \\
&\qquad + \tfrac{1}{2}\nu^{\mathrm{T}}\{\dot{V}_{\nu\nu} + V_{\nu x}f_uL_{uu}^{-1}f_u^{\mathrm{T}}V_{x\nu}\} = 0
\end{aligned} \tag{4.97}$$

which is of the form of Equation (4.93). V_{xx}, V_x, V_ν, $V_{x\nu}$, and $V_{\nu\nu}$ should be chosen so that

$$\begin{aligned}
-\dot{V}_{xx} &= L_{xx} + f_x^{\mathrm{T}}V_{xx} + V_{xx}f_x - V_{xx}f_uL_{uu}^{-1}f_u^{\mathrm{T}}V_{xx} \\
-\dot{V}_x &= f_x^{\mathrm{T}}V_x - V_{xx}f_uL_{uu}^{-1}f_u^{\mathrm{T}}V_x \\
-\dot{V}_\nu &= -V_xf_uL_{uu}^{-1}f_uV_{x\nu} \\
-\dot{V}_{x\nu} &= f_x^{\mathrm{T}}V_{x\nu} - V_{xx}f_uL_{uu}^{-1}f_u^{\mathrm{T}}V_{x\nu} \\
-\dot{V}_{\nu\nu} &= -V_{\nu x}f_uL_{uu}^{-1}f_u^{\mathrm{T}}V_{x\nu}
\end{aligned} \tag{4.98}$$

with the boundry conditions at the final time determined by Equation (4.90), i.e.,

$$V_{xx}(t_f) = \phi_{xx}, \quad V_x(t_f) = 0, \quad V_\nu(t_f) = -d, \quad V_{x\nu}(t_f) = M, \quad V_{\nu\nu}(t_f) = 0 \tag{4.99}$$

Now examining Equations (4.98) in the light of these boundary conditions, it can be seen that $V_x(t) = 0$ and $V_\nu(t) = -d$ for all t. Hence, the optimal choice for u, u^{opt} may be written

$$u^{\text{opt}} = -L_{uu}^{-1}(V_{xx}x + V_{x\nu}\nu) \tag{4.100}$$

The multiplier ν will now be eliminated from this expression by choosing ν^{opt} so that $V^{\text{ext}}(x, t)$ is extremal with respect to ν, i.e., from Equation (4.93)

$$\nu^{\text{opt}} = -V_{\nu\nu}^{-1}(V_{\nu x}x + V_\nu) \tag{4.101}$$

or

$$\nu^{\text{opt}} = -V_{\nu\nu}^{-1}(V_{\nu x}x - d) \tag{4.102}$$

Substituting this value for ν into Equations (4.100) and (4.93) gives

$$u^{\text{opt}}(x, t) = -L_{uu}^{-1} f_u^{\mathrm{T}} \{[V_{xx} - V_{x\nu} V_{\nu\nu}^{-1} V_{\nu x}]\, x + V_{x\nu} V_{\nu\nu}^{-1} d\} \tag{4.103}$$

and

$$V^{\text{opt}}(x, t) = V^0 - 1/2 d^{\mathrm{T}} V_{\nu\nu}^{-1} d - x V_{x\nu} V_{\nu\nu}^{-1} d + \tfrac{1}{2} x^{\mathrm{T}} [V_{xx} - V_{x\nu} V_{\nu\nu}^{-1} V_{\nu x}] x \tag{4.104}$$

which is the desired result. These equations describe the general feedback law and the associated return in terms of functions that may be obtained easily. As might be expected, the optimal control is a linear function of the state variables whereas the optimal return is a quadratic function of the state.

It must be noted, however, that these relations depend explicitly on the formation of the inverse of the matrix $V_{\nu\nu}$. Now, by definition $V_{\nu\nu}(t_f) = 0$, and hence these relations cannot be formed at the final time. In such regions, where it is not possible to form $V_{\nu\nu}^{-1}$, the system is said to be abnormal. Abnormality will occur if the terminal constraints are functionally interdependent or if the system is uncontrollable. Clearly, if $V_{\nu\nu}^{-1}$ does not exist at the initial time, the problem is ill formed and such situations must be avoided.

Along any given trajectory x^{opt}, ν^{opt} is a constant vector. Hence, ν^{opt} may be computed as soon as the matrix $V_{\nu\nu}$ becomes nonsingular by using Equation (4.102). The optimal control law is then

$$u^{\text{opt}}(t) = -L_{uu}^{-1} f_u^{\mathrm{T}} [V_{xx} x^{\text{opt}}(t) + V_{x\nu} \nu^{\text{opt}}] \tag{4.105}$$

However, it must be emphasized that this rule, Equation (4.105), applies only for the given trajectory x^{opt}, and, hence, is essentially *open loop*.

4.5.1 Example

EXAMPLE 4.4. Example 4.1 will be considered with the addition of a linear terminal constraint. The system equation is $\dot{x} = u$ and $x(0) = x_0$ and the performance index,

$$J = \int_0^T (x^2 + u^2)\, dt$$

The added constraint is $x(T) = 0$. This constraint is adjoined to the performance index with a multiplier ν to give

$$J^* = \nu x(T) + \int_0^T (x^2 + u^2)\, dt$$

The equation for V_{xx}, the Riccati equation, is unchanged, $-\dot{V}_{xx} = 1 - V_{xx}^2$ and the solution, in reverse time, is

$$V_{xx}(\tau) = \tanh(\tau) \tag{4.106}$$

The equation for $V_{x\nu}$ is $-\dot{V}_{x\nu} = -V_{xx}V_{x\nu}$, which in reverse time becomes $dV_{x\nu}/d\tau = -(\tanh \tau)V_{x\nu}$. Hence,

$$V_{x\nu}(\tau) = [\cosh(\tau)]^{-1} \tag{4.107}$$

Thus, the equation, in reverse time, for $V_{\nu\nu}$ is $dV_{\nu\nu}(\tau)/d\tau = [\cosh(\tau)]^{-2}$ and

$$V_{\nu\nu}(\tau) = \tanh(\tau) \tag{4.108}$$

Equations (4.106) and (4.108) specify the solution. Note that although $V_{\nu\nu}(\tau)$ is zero at the final time ($\tau = 0$), it could be inverted at any later time $\tau > 0$ obviating the need to continue the integration of the equations for $V_{x\nu}$ and $V_{\nu\nu}$.

4.6 STABILITY

In the foregoing analysis it was tacitly assumed that a finite solution to the matrix Riccati equation existed and that the resulting controlled system was stable. Unfortunately, it is very difficult to derive any general theorems about existance of the optimal control and stability, although if restrictions are placed on the system certain results can be obtained. The authors have encountered little difficulty with *practical* examples although, of course, it is very easy to construct artificial examples in which problems arise.

The reader interested in further study in this area should consult the references listed at the end of this chapter and the section on numerical stability in Chapter 8.

PROBLEMS

1. Extend the discussion of 4.3.3 to problems with $L(x, u, t) \neq 0$.
2. As in the discrete case the linear quadratic algorithm can be extended to include performance indices that contain linear terms, i.e., of the form

$$J = \tfrac{1}{2}x^{\mathrm{T}}(t_f)\,\phi_{xx}(t_f) + \phi_x x$$
$$+ \int_{t_0}^{t_f} \{\tfrac{1}{2}[x^{\mathrm{T}}L_{xx}x + x^{\mathrm{T}}L_{xu}u + u^{\mathrm{T}}L_{ux}x + u^{\mathrm{T}}L_{uu}u] + L_u u + L_x x\}\, dt$$

Show that the optimal control has the form

$$u(x, t) = -[D(t)\,x(t) + E(t)]$$

and that the optimal return has the form

$$V^{\text{opt}}(x(t_0), t_0) = \tfrac{1}{2}x^{\text{T}}(t_0)\,V_{xx}x(t_0) + V_x x(t_0) + V_0(t_0)$$

Obtain differential equations for V_{xx}, V_x, and V_0.

3. Find the control $u(x)$ that *minimizes* the performance criterion

$$J = \int_0^\infty \{x_1^2 + u^2\}\,dt$$

where the state equations are, $\dot{x}_1 = x_2 + u$ and $\dot{x}_2 = -u$.

Note: In this case the Riccati equation need not be integrated for $\dot{V}_{xx}(0) = 0$. Thus it is only necessary to solve the algebraic equation

$$0 = L_{xx} + V_{xx}f_x + f_x^{\text{T}}V_{xx} - V_{xx}f_u L_{uu}^{-1} f_u V_{xx}$$

4. Example 4.3 illustrated one technique for modeling a desired response. Another approach would be to choose x_8 by $\dot{x}_8 = \dot{x}_3 - x_{8/T}$ where T is a suitable time constant (e.g., here $T = 1$ sec). Specify the optimal control for Example 4.3 where the desired roll rate x_8 is given by the foregoing equation. What are the advantages and/or disadvantages of this approach?

5. Find V^{opt} and u^{opt} for Example 4.4 with the terminal constraint $x(t_f) = d$.

BIBLIOGRAPHY AND COMMENTS

Historically finite difference techniques were used by Bernoulli and Euler in the early development of the calculus of variations. A description of their approach is given in the following book.

Courant, R. and Robbins, H. (1956). *What is Mathematics.* Oxford Univ. Press, London and New York.

Section 4.3. For references to the development of the dynamic programming equations, see the bibliography at the end of Chapter 3.

Section 4.3.3. This approach is used by

Caratheodory, C. (1967). *Calculus of Variations and Partial Differential Equations of the First Order* (R. Dean, transl.). Holden-Day, San Francisco, California. (Originally published in German in 1935.)

Section 4.4. References to the linear quadratic algorithm were given at the end of Chapter 3.

Section 4.4.1. Further details on the theory and operation of gyroscopes may be found in books by:

Arnold, R. N. and Maunder, L. (1961). *Gyrodynamics and Its Engineering Applications.* Academic Press, New York.

Fernandez, M. and Macomber, G. R. (1962). *Inertial Guidance Engineering.* Prentice-Hall, Englewood Cliffs, New Jersey.

These airplane stabilization results were first reported by

Dyer, P., Noton, A. R. M., and Rutherford, D. (1966). "The Application of Dynamic Programming to the Design of Invariant Autostabilizers," *J. Roy. Aeronaut. Soc.*, Vol. 70, pp. 469–476.

There are many publications concerned with the application of the linear quadratic algorithm. For example, the interested reader may consult the following survey papers where further references are to be found.

Athans, M. (1966). "The Status of Optimal Control Theory and Applications for Deterministic Systems," *Trans. IEEE Autom. Control*, Vol. AC-11, pp. 580–596.

Bryson, A. E. (1967). "Applications of Optimal Control Theory in Aerospace Engineering," *J. Spacecraft and Rockets*, Vol. 4, p. 545.

Section 4.5. The development given here is similar to that used by

Dreyfus, S. E. (1967). "Control Problems with Linear Dynamics, Quadratic Criterion and Linear Terminal Constraints," *Trans. IEEE Autom. Control*, Vol. AC-12, pp. 323–324.

McReynolds, S. R. and Bryson, A. E. (1965). "A Successive Sweep Method for Solving Optimal Programming Problems," *Joint Autom. Control Conf.*, pp. 551–555.

Section 4.6. A much more detailed discussion of the stability of the matrix Riccati equation and of the controlled system is given by

Kalman, R. E. (1960). "Contributions to the Theory of Optimal Control," *Bol. Soc. Mat. Mex.*, Vol. 5, p. 102.

Lee, E. B. and Markus, L. (1967). *Foundations of Optimal Control Theory.* Wiley, New York.

5

The Gradient Method and the First Variation

In this chapter a gradient algorithm is developed and illustrated with several examples. The relationship of the algorithm to the first variation is examined and a stationarity condition for a local optimum is derived.

The gradient method is developed in two ways. First, the finite difference technique is used to extend the discrete gradient method developed in Chapter 3. Second, the algorithm is derived from the continuous dynamic programming equations discussed in the last chapter. The application of the algorithm is illustrated with several examples, two of which, an aircraft landing problem and a boiler control problem, represent fairly realistic situations.

The first variation is also developed in two ways, both by a dynamic programming and by a classical approach. It is shown how the condition that the first variation must vanish for every admissible variation leads to a stationarity condition for a local extremal solution.

5.1 THE GRADIENT ALGORITHM

The gradient algorithm for continuous problems is a direct method in which a sequence of nominal solutions, which converge to the optimum, is generated. The new nominal solution is generated by incrementing the old nominal in the direction of the *gradient* of the return function. Formally, the gradient method is easily derived from the gradient algorithm for discrete systems which was discussed in Chapter 3. The continuous system is replaced by its finite difference analog, and then as the mesh size tends to zero the discrete algorithm becomes the continuous algorithm.

First, consider the basic problem without control parameters and side constraints. The performance criterion J is given by

$$J = \phi(x(t), t) + \int_{t_0}^{t_f} L(x, u, t)\, dt \tag{5.1}$$

and the system equations are

$$\dot{x} = f(x(t), u(t), t), \qquad x(t_0) = x_0 \quad \text{specified} \tag{5.2}$$

The finite difference approximations to these equations are

$$J = \phi(x(N), t_N) + \sum_{i=0}^{N-1} \{L(x(i), u(i), t_i)\}\, h \tag{5.3}$$

and

$$x(k+1) = x(k) + h(f(x(k), u(k), t_k)) \tag{5.4}$$

where

$$x(k) = x(t_k), \qquad u(k) = u(t_k), \qquad t_k = kh + t_0 \tag{5.5}$$

and h is the mesh size.

For this problem the discrete gradient algorithm (Section 3.6.5) for computing the *new* nominal $u^{j+1}(i)$ becomes

$$u^{j+1}(i) = u^j(i) + \epsilon[V_x(i+1) f_u(i) + L_u(i)]^{\mathrm{T}} \tag{5.6}$$

where

$$V_x(N) = \phi_x(x(N), t_N) \tag{5.7}$$

$$V_x(i) = V_x(i+1) + h[V_x(i+1) f_x(i) + L_x(i)] \tag{5.8}$$

Now taking the limit as $h \to 0$, the algorithm becomes

$$u^{j+1}(t) = u^j(t) + \epsilon[V_x(t) f_u(t) + L_u(t)] \tag{5.9}$$

$$V_x(t_f) = \phi_x(x(t_f), t_f) \tag{5.10}$$

$$\dot{V}_x(t) = -V_x f_x - L_x \tag{5.11}$$

Thus the transition equation (5.8) becomes a first-order differential equation (5.11) which may be integrated *backward* in time using Equation (5.10) as the starting value for V_x.

5.2 THE GRADIENT ALGORITHM: A DYNAMIC PROGRAMMING APPROACH

Alternatively the algorithm may be derived directly from a consideration of continuous dynamic programming. It is assumed that a new nominal control function $u^{j+1}(t)$ has somehow been found for times greater than t_m. The return function associated with the control u^{j+1} is, at the time t_m, denoted by $V(x(t_m), t_m)$. The problem is to extend the new control u^{j+1} to the left of t_m while increasing the value of the return.

Suppose that at some time t_1 prior to t_m, the nominal control u^j is replaced by the new control u^{j+1} (the control u^{j+1} is used for *all* time

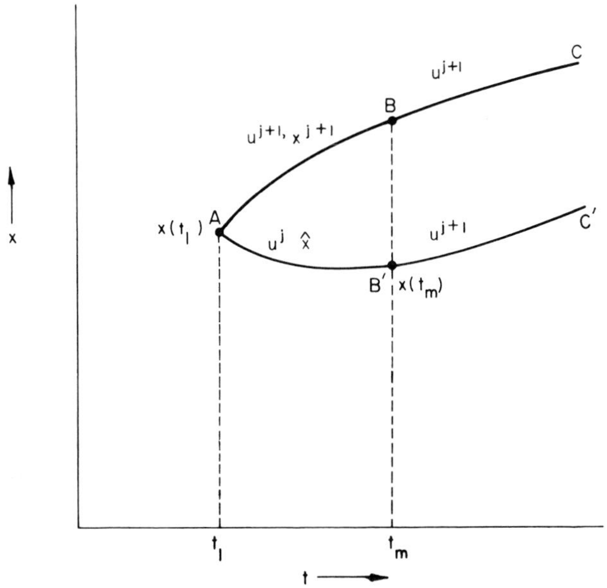

FIG. 5.1. The gradient algorithm.

after t_m), Figure 5.1. The difference in the return $\delta V(x_1, t_1)$ between the two paths (A, B, C and A, B', C') is given by

$$\delta V(x_1, t_1) = V(x_1, t_1) - \{V(\hat{x}(t_m), t_m) + \int_{t_1}^{t_m} L(\hat{x}(t), u^j, t)\, dt\} \qquad (5.12)$$

where $V(x_1, t_1)$ is the return associated with the control u^{j+1} at the time t_1 and where L is integrated along the trajectory $\hat{x}$ which is determined from

$$\dot{\hat{x}} = f(\hat{x}, u^j, t), \qquad \hat{x}(t_1) = x_1 \qquad (5.13)$$

Now x_1 and $\hat{x}$ do not necessarily coincide with the nominal state $x^j(t_1)$ and the nominal trajectory x^j, because of control changes *prior* to the time t_1. It is, however, assumed that x_1 and $\hat{x}$ lie in the *neighborhood* of $x^j(t_1)$ and x^j.

Next, the return $V(x_1, t_1)$ will be expressed in terms of an integral and $V(\hat{x}(t_m), t_m)$. The return $V(x, t)$ is equal to $V(\hat{x}(t_m), t_m)$ added to the integral of the rate of change of V along the path A, B', i.e.,

$$V(x_1, t_1) = \int_{t_m}^{t_1} \{V_x(\hat{x}, t) f(\hat{x}, u^j, t) + V_t(\hat{x}, t)\}\, dt + V(\hat{x}(t_m), t_m) \quad (5.14)$$

Hence, reversing the limits of the integral and substituting this expression for $V(x_1, t_1)$ into Equation (5.12) gives

$$\delta V(x_1, t_1) = -\int_{t_1}^{t_m} \{V_t(\hat{x}, t) + V_x(\hat{x}, t) f(\hat{x}, u^j, t) + L(\hat{x}, u^j, t)\}\, dt \quad (5.15)$$

Now the dynamic programming equation governing $V(\hat{x}, t)$ is [cf Equation (4.52)]

$$V_t + V_x f(\hat{x}, u^{j+1}, t) + L(\hat{x}, u^{j+1}, t) = 0 \quad (5.16)$$

Thus, eliminating V_t from the integral in Equation (5.15) by using Equation (5.16) gives

$$\delta V(x_1, t_1) = \int_{t_1}^{t_m} \{V_x[f(\hat{x}, u^{j+1}, t) - f(\hat{x}, u^j, t)] + L(\hat{x}, u^{j+1}, t) - L(\hat{x}, u^j, t)\}\, dt \quad (5.17)$$

Next, the right-hand side is expanded about the nominal control u^j to first order in δu

$$\delta V(x_1, t_1) \approx \int_{t_1}^{t_m} \{V_x f_u(\hat{x}, u^j, t) + L_u(\hat{x}, u^j, t)\}\, \delta u\, dt \quad (5.18)$$

Finally, this equation (5.18) is expanded about the nominal trajectory x^j to zeroth order in $\delta x = \hat{x} - x^j$, giving

$$\delta V(x^j, t_1) \approx \int_{t_1}^{t_m} \{V_x(x^j, t) f_u(x^j, u^j, t) + L_u(x^j, u^j, t)\}\, \delta u\, dt$$

Clearly δV will be positive (for a maximization problem) if δu is chosen by†

$$\delta u = \epsilon \operatorname{sgn}\{V_x f_u + L_u\} \quad (5.19)$$

where ϵ is a positive step length parameter that is chosen such that δu is small enough to ensure that the expansions are valid approximations.

† Obviously there are many ways in which δu could be chosen (see Section 5.2.1). However, Equation (5.19) represents the simplest form the algorithm takes.

It may be noted in passing that although the gradient expression for δu is independent of x^{j+1}, it does require that x^{j+1} is *close* to x^j.

It remains to determine $V_x(x^j, t)$. This is done by expanding Equation (5.16) about the nominal control u^j and the nominal trajectory x^j, viz,

$$0 = V_t(x^j, t) + V_x(x^j, t) f(x^j, u^j, t) + L(x^j, u^j, t)$$
$$+ [V_x{}^j f_u + L_u]\,\delta u + \left[\frac{dV_x}{dt} + V_x f_x + L_x\right]\delta x + 0(\delta x^2, \delta x\,\delta u, \delta u^2) \quad (5.20)$$

where all the partial derivatives are evaluated along the *nominal* trajectory. Since this expression must hold for arbitrary δx, to zeroth order in δu, V_x must satisfy the following differential equation

$$\dot{V}_x = -(V_x f_x + L_x) \quad (5.21)$$

and from the boundary condition $V(x, t_f) = \phi(x, t_f)$ we obtain,

$$V_x(x(t_f), t_f) = \phi_x(x(t_f), t_f) \quad (5.22)$$

Equations (5.19)–(5.22) complete the algorithm and are, of course, identical to Equations (5.9)–(5.11).

The gradient algorithm may now be summarized as follows.

1. Choose some nominal control u^j and integrate the state equations forward storing the state trajectory. Compute the cost J^j.

2. Using the stored values of the state x^j and control u^j compute V_x backward in time where

$$V_x(x(t_f), t_f) = \phi_x(x(t_f), t_f) \quad (5.23)$$

and

$$\dot{V}_x(t) = -V_x f_x(x^j, u^j, t) - L_x(x^j, u^j, t) \quad (5.24)$$

At the same time, compute *and* store the gradient $V_x f_u{}^j + L_u{}^j$.

3. Integrate the state equations forward using the new control u^{j+1} where

$$u^{j+1} = u^j + \epsilon \operatorname{sgn}(V_x f_u{}^j + L_u{}^j) \quad (5.25)$$

and where ϵ is some positive constant. Compute the new cost J^{j+1}.

4. If $J^{j+1} > J^j$, repeat step 2 with $j = j + 1$. If $J^j > J^{j+1}$, reduce the step size ϵ and repeat step 3.

5. Steps 2, 3, and 4 are repeated until the optimal solution has been obtained or until no further improvement can be obtained.

5.2.1 Choice of the Step Length ϵ

It is perhaps relevant to comment here that the main drawback to this procedure is the choice of the step-length parameter ϵ. For, if the parameter is too small the convergence is extremely slow, and if it is too large, convergence may not be obtained.

Indeed there is no reason for ϵ to be a constant; it could be chosen as a function of time. Perhaps the most common technique is to choose the step length to be proportional to the magnitude of the gradient so that δu becomes

$$\delta u = \epsilon' \{V_x f_u + L_u\} \tag{5.26}$$

In this case, the algorithm is usually referred to as the *method of steepest ascent*. However, a step-length parameter ϵ' must still be chosen. It is the authors' experience that, in general, an automatic procedure for choosing the step length must be built into the computer program if convergence is to be assured. For example, one idea is to increase the step size after each successful iteration by a factor η and to reduce it after an unsuccessful iteration by a factor β. Values for η in the range $2 \to 10$ and values of β, $0.1 \to 0.5$, have been found to be very successful.

One further idea, which is an attempt to reduce the variation in step size, is based on a rather naive second-order expansion in u of Equation (5.17), viz,

$$\delta V \approx \int_{t_1}^{t_m} \{(V_x f_u + L_u)\, \delta u + \tfrac{1}{2} \delta u^{\mathrm{T}} (V_x f_{uu} + L_{uu})\, \delta u\}\, dt \tag{5.27}$$

Provided the matrix $V_x f_{uu} + L_{uu}$ is negative definite, δV will be large as possible if δu is chosen by

$$\delta u = -(V_x f_{uu} + L_{uu})^{-1} (V_x f_u + L_u)^{\mathrm{T}}$$

Note, however, that this correction policy does not account for changes in the nominal state due to previous changes in the control. The full second-order expansion forms the basis of the successive sweep method which is described in the next chapter.

5.2.2 Control Parameters

In this and the following three sections the modifications needed to include control parameters, side constraints, and free terminal time will be discussed.

If control parameters α are present, some additional steps must be included. The return is expanded about some nominal control parameter α^j at the *initial time*, i.e.,

$$V(x(t_0), \alpha^{j+1}, t_0) \approx V(x(t_0), \alpha^j, t_0) + V_\alpha{}^j \delta\alpha + \cdots \tag{5.28}$$

Clearly, to ensure that $V(\alpha^{j+1}) > V(\alpha^j)$, $\delta\alpha = \alpha^{j+1} - \alpha^j$ should be chosen by $\delta\alpha = \epsilon[V_\alpha{}^j(t_0)]$ where ϵ is some positive constant that is chosen small enough to ensure the validity of the expansion. A differential equation for V_α is easily found by extending the expansion of Equation (5.16) given by Equation (5.20) to include terms in α, i.e.,

$$\begin{aligned} 0 \approx{} & V_t(x^j, \alpha^j, t) + V_x(x^j, \alpha^j t) f(x^j, u^j, \alpha^j) + L(x^j, u^j, \alpha^i, k) \\ & + [V_x{}^j f_u + L_u]\,\delta u + \left[\frac{dV_x{}^j}{dt} + V_x f_x + L_x\right] \delta x \\ & + \left[\frac{dV_\alpha{}^j}{dt} + V_x f_\alpha + L_\alpha\right] \delta\alpha + \cdots \end{aligned}$$

Hence,

$$V_\alpha{}^j(t_f) = \phi_\alpha, \qquad \dot{V}_\alpha = -V_x f_\alpha - L_\alpha \tag{5.29}$$

5.2.3 Initial Conditions

It has been assumed that the initial state $x(t_0)$ was specified. This may not always be so, in which case the gradient correction $\delta x(t_0)$ is given by

$$\delta x(t_0) = x^{j+1}(t_0) - x^j(t_0) = \epsilon[V_x(x(t_0), t_0)]^T \tag{5.30}$$

where ϵ is again a positive parameter used to control the size of $\delta x(t_0)$.

5.2.4 Side Constraints

If side constraints are present, further modifications to the basic procedure must be made. For example, consider a side constraint of the form [cf Equation (4.6)]

$$K = \psi(x(t_f), \alpha) + \int_{t_0}^{t_f} \{M(x(t), \alpha, u, t)\}\, dt \tag{5.31}$$

As in the discrete problem a gradient projection procedure is used (cf Section 3.6). Thus a new return function $W(x(t), \alpha, t)$ is introduced where

$$W(x, \alpha, t_0) = \psi(x(t_f), \alpha) + \int_{t_0}^{t_f} \{M(x, \alpha, u, s)\}\, ds \tag{5.32}$$

Now, following a development similar to that just given, it is possible to form the gradient of W, $W_x f_u + M_u$ where W_x is given by

$$W_x(x(t_f), t_f) = \psi(x(t_f), \alpha), \qquad \dot{W}_x = -W_x f_x - M_x \tag{5.33}$$

The modified algorithm becomes (cf Section 3.6.1)

$$u^{j+1} = u^j + \epsilon[V_x f_u + L_u + \nu^{\mathrm{T}}(W_x f_u + M_u)]^{\mathrm{T}} \tag{5.34}$$

and

$$\alpha^{j+1} = \alpha^j + \epsilon[V_\alpha(t_0) + \nu^{\mathrm{T}} W_\alpha(t_0)]^{\mathrm{T}} \tag{5.35}$$

It remains to choose values for the parameters ν. They are chosen to enforce a predetermined change δK in the level of the linearized constraint. In terms of the return W, δK is given to first order by

$$\delta K = W_\alpha(t_0)\,\delta\alpha + \int_{t_0}^{t_f} (W_x f_u + M_u)\,\delta u\, dt \tag{5.36}$$

Substituting $\delta\alpha$ and δu from Equations (5.34) and (5.35) gives

$$\delta K = \epsilon W_\alpha(t_0)(W_\alpha^{\mathrm{T}}(t_0)\nu + V_\alpha^{\mathrm{T}}(t_0)]$$
$$+ \epsilon \int_{t_0}^{t_f} (W_x f_u + M_u)[(W_x f_u + M_u)^{\mathrm{T}}\nu + V_x f_u + L_u]\, dt$$

Hence, ν is given by

$$\nu = \left[\epsilon\left(\| W_\alpha(t_0)\|^2 + \int_{t_0}^{t_f} \| W_x f_u + M_u \|^2\, dt\right)\right]^{-1}$$
$$\times \left\{\delta K - \epsilon\left[W_\alpha(t_0)\, V_\alpha^{\mathrm{T}}(t_0) + \int_{t_0}^{t_f} (W_x f_u + M_u)(V_x f_u + L_u)^{\mathrm{T}}\, dt\right]\right\} \tag{5.37}$$

Of course it is assumed that the condition

$$\| W_\alpha(t_0)\|^2 + \int_{t_0}^{t_f} \| W_x f_u + M_u \|^2\, dt = 0 \tag{5.38}$$

does not occur. For this condition would imply that $W_\alpha(t_0) = 0$, $W_x f_u + M_u = 0$, and $t_0 \leqslant t \leqslant t_f$ or, in other words, that the control parameters and control variables cannot change the side constraints *to first order*. If more than one side constraint is present, the function

$$\| W_\alpha \|^2 + \int_{t_0}^{t_f} \| W_x f_u + M_u \|^2\, dt$$

is a matrix. If this matrix is singular, then a vector σ must exist such that

$$0 = \sigma^{\mathrm{T}} W_{\alpha}, \qquad 0 = \sigma^{\mathrm{T}}(W_{\alpha} f_u + M_u) \tag{5.39}$$

This would imply that, at least to first order, independent changes in the constraint levels are not possible. Both of these situations are cases of *abnormality*. Abnormality will occur if the side constraints are functionally interdependant or if the system is uncontrollable. Such situations indicate an ill-formed problem and obviously must be avoided.

5.2.5 Free Terminal Time

One further class of problems that will be considered is characterized by a free terminal time. This type of problem is interesting because *strong* variations have to be employed.

Consider a change in the final time from t_f to $t_f + dt_f$. The return $V(x(t_f + dt_f), t_f + dt_f)$ at the final time is given by

$$V(x(t_f + dt_f), t_f + dt_f) = \Phi(x(t_f + dt_f), t_f + dt_f) + \int_{t_f}^{t_f+dt_f} L(x, u, t)\, dt \quad (5.40)^{\dagger}$$

Now the right-hand side is expanded about the nominal time t_f to *first order*,

$$V(t_f + dt_f) = \Phi(t_f) + [L(x, u, t_f) + \Phi_t]\, dt_f + \Phi_x \Delta x \tag{5.41}$$

where Δx (see Figure 5.2) is given, to first order in dt_f, by $\Delta x = x(t_f + dt_f) - x(t_f) \approx f\, dt_f$ and where all the partial derivatives are evaluated at the nominal t_f. Eliminating Δx, Equation (5.41) becomes

$$V(t_f + dt_f) = \Phi(t_f) + [L(x, u, t_f) + \Phi_t + \Phi_x f]\, dt_f + \cdots \tag{5.42}$$

Next, consider the effects of variations in the nominal state $x^j(t_f)$, $\delta x(t_f)$, and V becomes, to first order,

$$V(t_f + dt_f) \approx \Phi(x^j(t_f), t_f) + [L(x^j, u, t_f) + \Phi_t + \Phi_x f]\, dt_f + \Phi_x\, \delta x \cdots \tag{5.43}$$

Thus, to ensure that $V(x(t_j + dt_f), t_f + dt_f) > V(x(t_f), t_f)$ (for a maximization problem), the change in the final time dt_f must be chosen by

$$dt_f = \epsilon \operatorname{sgn}\{L(x, u, t_f) + \Phi_t + \Phi_x f\} \tag{5.44}$$

where ϵ is some positive constant chosen to ensure the validity of the expansion. The modifications to the basic algorithm are as follows.

† $\Phi(x(t_f), t_f)$ here is defined by $\Phi = \phi + \nu\psi^{\mathrm{T}}$ where ν is the multiplier determined by Equation (5.37). If there are no side constraints, $\Phi = \phi$.

1. When integrating the nominal state x^j forward, the integration is stopped at the nominal time t_f^j.

2. At the final time t_f^j, the change in the nominal time dt_f is computed from Equation (5.44).

3. When integrating the *new* nominal state x^{j+1} forward, the integration is stopped at the *new* nominal time t_f^{j+1} where $t_f^{j+1} = t_f^j + dt_f$.

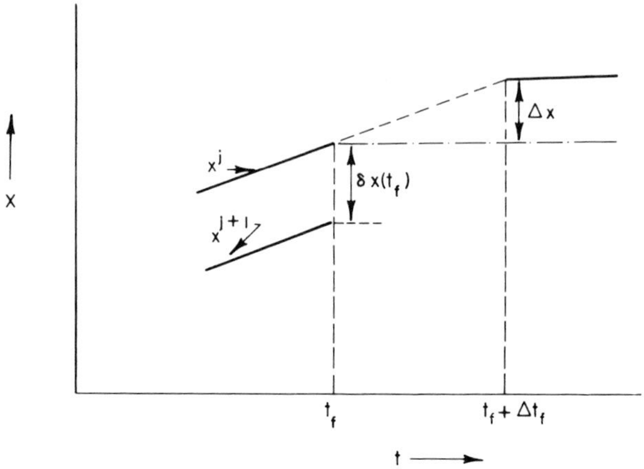

FIG. 5.2. Changes in the final time.

Another convenient technique for handling possible changes in the terminal time relies on employing one of the terminal constraints as a *stopping condition*, i.e., the nominal trajectory is integrated forward until this constraint is satisfied.

Suppose that at the jth iteration this terminal constraint is denoted by Ω^j, i.e.,

$$\Omega^j(x(t_f), t_f) = 0 \tag{5.45}$$

The change in Ω, from the jth to the $j+1$st iteration, is given to first order by

$$\Omega^{j+1} - \Omega^j = \delta\Omega = (\Omega_x{}^j f + \Omega_t{}^j)\, dt_f + \Omega_x{}^j\, \delta x\,(t_f) \tag{5.46}$$

where the partial derivatives are evaluated at $x^j(t_f)$, $t_f{}^j$. Hence it is possible to solve for dt_f in terms of $\delta\Omega$ and δx, i.e.,

$$dt_f = [\delta\Omega - \Omega_x\, \delta x]/\dot{\Omega} \qquad (5.47)^\dagger$$

† In general, $\Omega^{j+1} = \Omega^j = \delta\Omega = 0$. However, it may not always be possible to satisfy this condition on the first few iterations, and hence $\delta\Omega$ may not be zero.

By using this expression to eliminate dt_f from Equation (5.43), we obtain

$$V(t_f + dt_f) = [\phi_x - (L + \Phi_t + \Phi_x f)\, \Omega_x/\dot{\Omega}]\, \delta x + [L + \Phi_t + \Phi_x f]\, \delta\Omega/\dot{\Omega} \quad (5.48)$$

and thus it is necessary only to modify the partial derivative of V at the final time, i.e.,

$$V_x(t_f^{j+1}) = [\phi_x - (L + \Phi_t + \Phi_x f)\, \Omega_x/\dot{\Omega}] \quad (5.49)$$

Note that this technique requires the inversion of $\dot{\Omega}$. The modifications to the algorithm then become:

1. When integrating the nominal state forward the integration is stopped when the stopping condition is satisfied.
2. The initial condition in the backward integration for V_x is given by Equation (5.49).

5.3 EXAMPLES

EXAMPLE 5.1. First, as a simple example, consider the minimization of the performance index J where $J = \frac{1}{2} \int_0^1 (x_1^2 + u^2)\, dt$ and where the system equations are $\dot{x}_1 = x_2 + u$, $x_1(0) = 2$, $\dot{x}_2 = -u$, and $x_2(0) = -1$.

The partial derivatives required to utilize the gradient algorithm are $L_x = [x_1, 0]$, $L_u = [u]$, $f_x = \begin{bmatrix} 0 & 1 \\ 0 & 0 \end{bmatrix}$, and $f_u = \begin{bmatrix} 1 \\ -1 \end{bmatrix}$. The boundary conditions for the partial derivatives V_{x_1}, V_{x_2} are $V_{x_1}(1) = 0$ and $V_{x_2}(1) = 0$.

The nominal control u^j is chosen as $u^j(t) = 0$ and $0 \leqslant t \leqslant 1$.

The gradient algorithm is then used to compute the optimal control. The convergence of the cost is shown in Table XVI. As the example was linear quadratic, the true optimal control is computed easily and is also shown in the table. It should be noted that, in spite of the simplicity of the example, three gradient steps were required to obtain the optimal control whereas the linear quadratic algorithm obtained the solution immediately.

TABLE XVI

CONVERGENCE OF COST FOR EXAMPLE 5.1

Nominal	1.16666
1	1.05098
2	1.04886
3	1.04874
Optimal	1.04874

Still, the gradient algorithm appears to provide fairly fast convergence toward the optimum. This rapid convergence is *not* a characteristic of the gradient method, as will be seen in the following examples.

EXAMPLE 5.2. Next, to illustrate the convergence when the system is nonlinear, the brachistochrone problem (Example 3.6.3) will be considered. This example also illustrates the application of the gradient projection procedure. As the problem was described in detail in Chapter 3, only the relevant equations will be repeated here.

The problem is to minimize a performance index J where

$$J = \int_0^1 [1 + u(t)^2]^{1/2}/[1 - x(t)]^{1/2} \, dt \tag{5.50}$$

subject to the terminal constraint

$$\psi = x(1) + 0.5 - 0 \tag{5.51}$$

and with the state x governed by the equation

$$\dot{x} = u, \qquad x(0) = 0 \tag{5.52}$$

The partial derivatives needed to implement the gradient algorithm are

$$\begin{aligned} L_x &= 0.5(1 + u^2)^{1/2}/(1 - x)^{3/2} \\ L_u &= u/\{(1 + u^2)^{1/2}(1 - x)^{1/2}\} \\ f_x &= 0 \qquad \text{and} \qquad f_u = 1 \end{aligned}$$

Thus, the characteristic equation for V_x becomes

$$\dot{V}_x = -0.5(1 + u^2)^{1/2}/(1 - x)^{3/2}, \qquad V_x(1) = 0 \tag{5.53}$$

and for W_x becomes

$$\dot{W}_x = 0, \qquad W_x(1) = 1 \tag{5.54}$$

The change in the control δu is given by

$$\delta u = \epsilon\{V_x + u/[(1 + u^2)^{1/2}(1 - x)^{1/2}] + \nu\} \tag{5.55}$$

where ν is given by [cf Equation (5.37)]

$$\nu = \delta K/\epsilon - \int_0^1 (\delta u/\epsilon) \, dt \tag{5.56}$$

(Note that as this is a *minimization* problem, ϵ will be negative. Here a value of -2 was used.) The change in the control level δK was chosen to be the error in the terminal constraint, i.e., $\delta K = -(x(1) + 0.5)$. This completes the relations needed for the algorithm. The procedure used in this example may now be summarized as follows.

1. The state equation (5.52) was integrated forward using the nominal control $u(s) = 0$, $0 \leqslant s \leqslant 0.5$, $u(s) = -1$, $0.5 \leqslant s \leqslant 1$. The state was stored at each integration step and the cost was also computed.

2. The equation for V_x, Equation (5.53), was integrated backward in time using the stored values of the state and the nominal control. At the same time the gradient $V_x + u/[(1 + u^2)^{1/2}(1 - x)^{1/2}]$ was computed and stored.

3. The parameter ν was computed using Equation (5.56).

4. The correction δu was computed using Equation (5.55).

5. The state equation was again integrated forward using the new nominal control u^{j+1} where $u^{j+1} = u^j + \delta u$.

6. Steps 2–5 were repeated until no further reduction in the cost was obtained.

The convergence is shown in Table XVII and should be compared with Table XIX for the discrete case. The state and control histories are

TABLE XVII. BRACHISTOCHRONE PROBLEM

Iteration Number	Cost	$u(0)$	$x(1)$	ν
Nominal	1.1356745	−0.0000000	−0.5000000	−0.0000000
1	1.0236765	−1.1497389	−0.5000000	0.0653564
2	1.0000128	−0.7794464	−0.5000000	0.1794485
3	0.9986935	−0.7902167	−0.5000000	0.2188919
4	0.9985214	−0.7796591	−0.5000000	0.2185983
5	0.9985025	−0.7843331	−0.5000000	0.2195242
6	0.9984992	−0.7827871	−0.5000000	0.2192970
7	0.9984990	−0.7833571	−0.5000000	0.2193884
8	0.9984988	−0.7831475	−0.5000000	0.2193571
9	0.9984988	−0.7832273	−0.5000000	0.2193689

very similar to those for the discrete case (Figure 3.6) and so are not repeated here.

The equations were integrated with a standard fourth-order Runge–Kutta scheme with a step length of 0.01. On the backward integration the stored values of the state were linearly interpolated where necessary.

The computer program took approximately 10 sec of IBM 7090 computer time including printout.

EXAMPLE 5.3. Another version of the brachistochrone problem will now be considered to illustrate the effects of a variable final time. The state equation (5.52) remains unaltered but the performance criterion becomes

$$J = \int_0^{t_f} \{(1 + u^2)^{1/2}/(1 - x)^{1/2}\}\, dt \tag{5.57}$$

where the final time t_f is free. Also, the terminal constraint is changed to

$$\psi_1 = x(t_f) + 0.5 - t_f = 0 \tag{5.58}$$

First this constraint will be used as a stopping condition to eliminate variations in the final time. Thus, the correction δu to the nominal control is now given by $\delta u = \epsilon\{V_x f_u + L_u\}$. The equation for V_x is the same except that the boundary condition $V_x(t_f)$ is given by [cf Equation (5.49)]

$$V_x(t_f) = [1 + u(t_f)^2]^{1/2}/\{[1 - x(t_f)]^{1/2}[u(t_f) - 1]\}$$

The procedure may now be summarized briefly as follows.

1. The state equations are integrated forward until the stopping condition equation (5.58) is satisfied, using the nominal control $u^j = -1$.

2. The equation for V_x is integrated backward using Equation (5.58) as the starting condition.

3. The state equation is integrated forward using the new nominal u^{j+1} where $u^{j+1} = u^j + \epsilon\{V_x f_u + L_u\}$ and where $\epsilon = -2$.

4. Steps 1, 2, and 3 were repeated until convergence was obtained.

A summary of the convergence is given in Table XVIII. The state and control trajectories are shown in Figures 5.3a and b. As can be seen from Table XVIII, 16 iterations are required and the convergence is somewhat slower than in the previous example. This, of course, is to be expected as the free terminal time represents a severe nonlinearity.

As a general rule this stopping condition technique should be used if at all possible. The priming trajectory does *not* have to satisfy the condition exactly, although it does help convergence. In cases where the constraint is not satisfied exactly the forward integration should be stopped when the constraint is most nearly satisfied. The main advantages of the method are that it is simple to program, it eliminates one of the side constraints, and, furthermore, it avoids the necessity of choosing a step length when forming the new nominal final time t_f.

TABLE XVIII. CONVERGENCE FOR EXAMPLE 5.3

	Cost	$u(0)$	$x(t_f)$	t_f
Prime	0.3338505	−1.0000000	−0.2500000	0.2500000
1	0.3332942	−1.1493023	−0.2591621	0.2408379
2	0.3332432	−1.1974557	−0.2618201	0.2381799
3	0.3332371	−1.2175405	−0.2327743	0.2372257
4	0.3332366	−1.2265593	−0.2631393	0.2368607
5	0.3332367	−1.2307483	−0.2632837	0.2367163
6	0.3332368	−1.2327302	−0.2633421	0.2366579
7	0.3332369	−1.2341357	−0.2633763	0.2366237
8	0.3332369	−1.2341357	−0.2633763	0.2366237
9	0.3332369	−1.2343573	−0.2633824	0.2366176
10	0.3332369	−1.2344651	−0.2633824	0.2366176
11	0.3332369	−1.2345178	−0.2633824	0.2366168
12	0.3332369	−1.2345435	−0.2633836	0.2366164
13	0.3332369	−1.2345562	−0.2633838	0.2366162
14	0.3332369	−1.2345624	−0.2633838	0.2366162
15	0.3332369	−1.2345654	−0.2633838	0.2366162
16	0.3332369	−1.2345669	−0.2633838	0.2366161

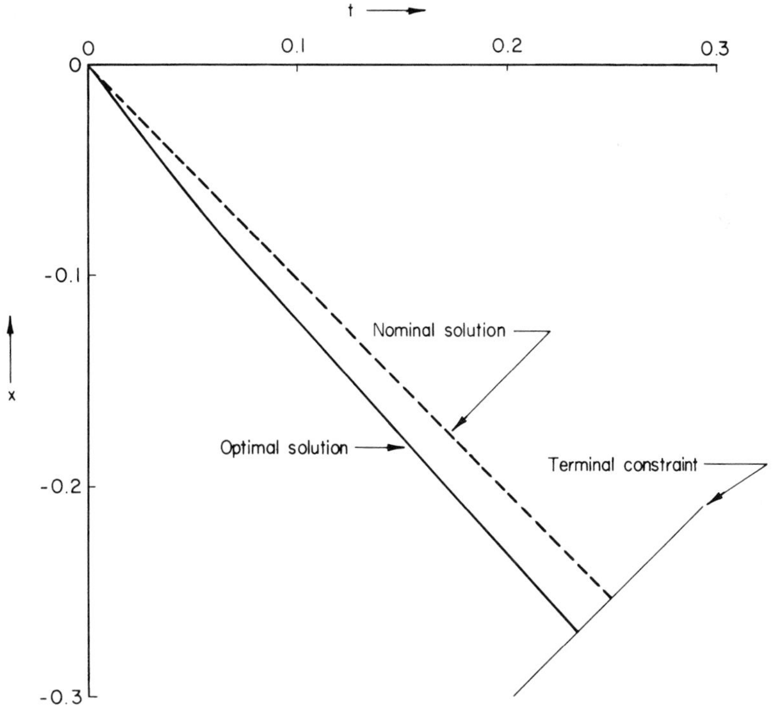

FIG. 5.3a. Optimal state trajectory (Example 5.3).

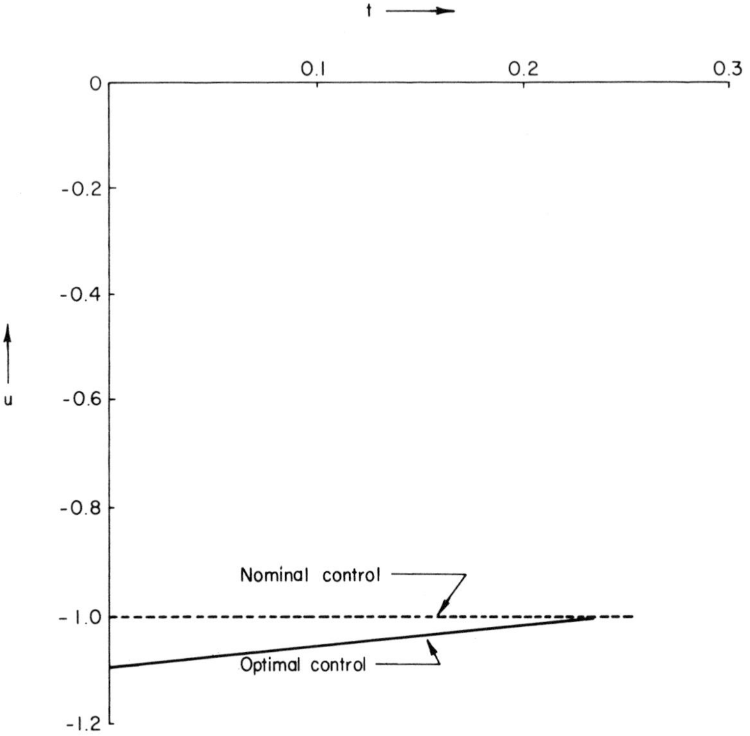

FIG. 5.3b. Optimal control trajectory (Example 5.3).

However, if the technique cannot be used, e.g., if $\dot{\psi} = 0$, changes in the final time may be computed from Equation (5.44). In this example the correction dt_f is given by

$$dt_f = \epsilon\{(1 + u(t_f)^2)^{1/2}/(1 - x(t_f))^{1/2} + \nu(u(t_f) - 1)\}$$

(**Note:** ϵ is negative as this is a minimization problem.)

The procedure now closely follows that of Example 5.2 (except for the different terminal constraint) except that the state equations are integrated forward to the *nominal* final time t_f^j. Also, at each iteration a new nominal time t_f^{j+1} is formed using $t_f^{j+1} = t_f^j + \epsilon\, dt_f$. The step length ϵ was chosen to be -0.1.

The convergence is shown in Table XIX. Note that convergence required 28 iterations, considerably more than with the previous technique. It might be hoped that increasing the step length ϵ would speed up the convergence, but in practice the procedure tended to diverge. Thus, this method, at least in this case, is much slower.

TABLE XIX. CONVERGENCE FOR EXAMPLE 5.3

Cost	$u(0)$	$x(1)$	ν	T
0.3338505	−1.0000000	−0.2500000	−0.0000000	0.2500000
0.3247017	−1.0758186	−0.2431542	0.5957137	0.2426516
0.3318283	−1.1584941	−0.2561851	0.6283643	0.2414487
0.3305803	−1.1799180	−0.2564389	0.6199097	0.23923701
0.3322798	−1.2033219	−0.2600442	0.6273728	0.2385766
0.3322388	−1.2125542	−0.2606302	0.6259198	0.2377771
0.3328167	−1.2211752	−0.2618892	0.6281090	0.2374425
0.3328363	−1.2252849	−0.2622396	0.6278778	0.2371246
0.3330414	−1.2287465	−0.2627205	0.6286085	0.2369700
0.3330710	−1.2305849	−0.2628985	0.6286085	0.2369700
0.3331481	−1.2320267	−0.2630908	0.6288682	0.2367687
0.3331670	−1.2328463	−0.2631759	0.6288956	0.2367136
0.3331971	−1.2334577	−0.2632548	0.6289932	0.2367136
0.3332072	−1.2338214	−0.2632942	0.6290134	0.2366588
0.3332192	−1.2340832	−0.2633272	0.6290512	0.2366448
0.3332242	−1.2342438	−0.2633451	0.6290625	0.2366348
0.3332291	−1.2343565	−0.2633590	0.6290775	0.2366286
0.3332314	−1.2344272	−0.2633670	0.6290833	0.2366243
0.3332335	−1.2344759	−0.2633766	0.6290921	0.2366197
0.3332346	−1.2345070	−0.2633791	0.6290946	0.2366185
0.3332354	−1.2345281	−0.2633791	0.6290946	0.2366185
0.3332359	−1.2345417	−0.2633807	0.6290959	0.2366177
0.3332363	−1.2345508	−0.2633818	0.6290969	0.2366172
0.3332365	−1.2345568	−0.2633825	0.6290975	0.2366168
0.3332366	−1.2345607	−0.2633830	0.6290980	0.2366166
0.3332367	−1.2345633	−0.2633833	0.6290982	0.2366164
0.3332368	−1.2345651	−0.2633835	0.6290984	0.2366163
0.3332368	−1.2345662	−0.2633836	0.6290985	0.2366163
0.3332369	−1.2345670	−0.2633837	0.6290986	0.2366162
0.3332369	−1.2345674	−0.2633837	0.6290987	0.2366162
0.3332369	−1.2345678	−0.2633838	0.6290987	0.2366162
0.3332369	−1.2345680	−0.2633838	0.6290987	0.2366162
0.3332369	−1.2345681	−0.2633838	0.6290987	0.2366162
0.3332369	−1.2345682	−0.2633838	0.6290988	0.2366161

The computer program details for this problem are essentially the same as for those in Example 5.2. In particular, the step length used in updating the nominal control was kept at −2.

EXAMPLE 5.4. *An Aircraft Landing Problem.* Next, a somewhat more complicated example will be discussed, viz, the automatic landing of an aircraft. In this study only the final stage of the landing, the flareout phase, will be considered. This maneuver begins when the aircraft is about 100 ft above the ground. It is assumed that the aircraft will be guided to an acceptable location for initiating the flare by some existing guidance system. Also, only the forward motion, the *longitudinal mode* is considered, because the function of the lateral controls is merely to point the aircraft down the runway.

The first step in defining the optimal control problem is to set up a suitable performance criterion. Unfortunately, and this is a characteristic of many practical problems, it is extremely difficult to specify all of the features that determine a satisfactory performance index.

Here it is assumed that at least following requirements must be satisfied.

1. The rate of descent $\dot{h}$ should remain constant for the last few seconds before touchdown, and at touchdown should be between 2.0 and 0.5 ft/sec, i.e., the aircraft must not "float" over the runway. Clearly, if the aircraft floated over the runway at a height of a few feet, the effects of gusts and turbulence would be accentuated, and might, for example, cause the aircraft to overshoot the runway or to stall.

2. Although the forward speed V should drop during the landing phase, it must remain higher than the stall speed V_s. In practice, to ensure stability in the presence of gusts the speed is maintained at least ten per cent above the stall speed.

3. The pitch angle θ at touchdown must be positive so that the nosewheel does not touch the runway before the main undercarriage.

4. The pitch rate $\dot{\theta}$ must be less than 1.5 deg/sec so that there are no sudden changes in pitch angle. This is particularly important in passenger aircraft if passenger comfort is to be ensured. It is also important in the last few seconds before touchdown, in particular, a large pitch rate should not be required to meet condition 5.

5. The elevator deflection η must be less than 20 deg, and the elevator rate $\dot{\eta}$ must be less than 6 deg/sec.

6. The throttle setting Th should lie between 0 and 1 (fully closed).

7. Finally, the landing should take a maximum of 20 sec.

It must be emphasized that the foregoing requirements are in no way exhaustive. Also, the numbers used (e.g., maximum landing time—20 sec) depend on the aircraft considered as well as on many other considerations. However, changes in the numerical values will not affect the basic procedure. It is interesting to note, as might be expected, that some of these requirements are conflicting. For example, any change that tends to increase the pitch angle will, in general, tend to reduce the forward speed (i.e., increase the tendency to stall) and vice versa.

These requirements lead to a choice of performance criterion of the following form.

$$V = \phi_T + \int_0^{20} [A_1(u_A - u_D)^2 + A_2(h - h_D)^2 + g_1\dot{\eta}^2 + g_2(Th + c)^{2m}]\, dt \quad (5.59)$$

where u_A equals the change in forward speed (feet per second); u_D equals the desired change in forward speed (feet per second); h_D is desired height trajectory (feet); and $\dot{\theta}$ represents the pitch rate (degrees per second).

The parameters A_1, A_2, g_1, g_2, c, and m control the relative weightings of the terms and are determined later. The function ϕ_T represents the terminal conditions.

The desired change in forward speed trajectory u_D was determined by assuming that the speed at touchdown should be $1.1 \times V_s$. Hence, u_D is given by $u_D = [V_A - 1.1\, V_S]t/T$ where t is the time from the start of flare (seconds) and V_A is the approach speed (feet per second). Providing the aircraft follows this trajectory, the pitch angle will be positive at touchdown without the forward speed, even in the presence of gusts, dropping below the stall speed.

The choice of a height trajectory h_D is somewhat arbitrary and is, to some extent, a question of personal opinion.† In this case, the one suggested by Merriam (1964) was used, viz,

$$h_D = 100 \exp(-t/5), \qquad t < 15$$

$$h_D = 20 - t, \qquad t \geqslant 15$$

The rate of descent $\dot{h}_D$ is constant for the last 5 sec and at touchdown is equal to 1.0 ft/sec, and so the trajectory satisfies condition 1.

The first two terms in the performance criterion weight deviations from these trajectories; hence, the aircraft will satisfy conditions 1 and 2. It was felt that it would be unnecessary to emphasize the landing trajectory with a weighting on height at the terminal time. The most likely effect of such a term would be a sharp reduction of the pitch angle unless the aircraft was on the correct trajectory, and, in any case, small variations in the landing time could be tolerated.

The third term weights the pitch rate (condition 4) and helps promote a smooth landing. The remaining two terms control the amplitude of the forcing signals $\dot{\eta}_D$ and Th. The first of these terms weights the elevator rate (condition 6), and the second the throttle setting.

Finally, a separate weighting was put on the pitch angle at the final time to ensure that condition 3 was met, viz $\phi_T = A\theta(T)^2$.

† In fact, it might be better to eliminate the idea of a reference trajectory altogether.

The next step in setting up the problem is to determine the equations of motion or state equations. The longitudinal aerodynamics are described by four equations of motion that are very similar to the equations for level flight. The main differences are that the equations have been linearized about the glide path rather than level flight, and nonlinear terms have been included to account for ground effects. Three more equations are required to include height and actuators. Figure 5.4

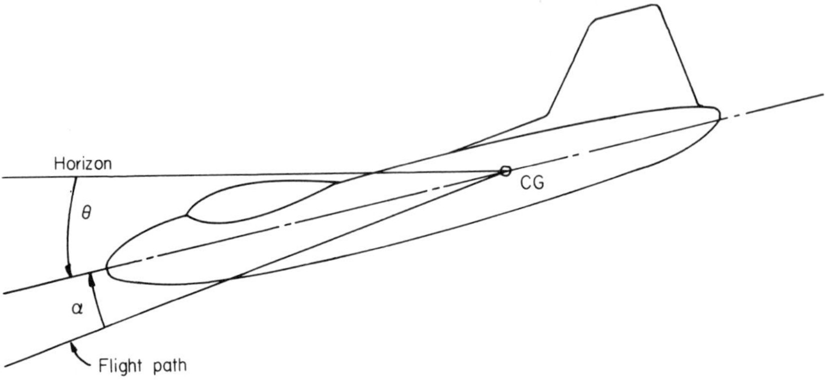

FIG. 5.4. Definition of angles.

illustrates the overall pitch, θ_{100}, and attitude, α_{100}, angles, and the forward speed at 100 ft. The state variables in the following equations represent *deviations* from these angles and this speed, e.g., $\theta = \theta_{\text{at height } h} - \theta_{100}$.

Typical state equations for a heavy transport aircraft during flare (aircraft weight, 215,000 lb, approach speed, 232 ft/sec) are

$$\begin{aligned}
\dot{x}_1 &= -0.041x_1 + 0.381x_2 - 0.562x_3 - 2.552x_7 + 0.221R \\
\dot{x}_2 &= -0.066x_1 - 0.577x_2 + x_3 - 0.050x_6 - 0.992x_7 - 0.395R \\
\dot{x}_3 &= 0.011x_1 - 1.108x_2 - 0.822x_3 - 1.264x_6 - .157x_7 - 3.544R \\
\dot{x}_4 &= x_3 \\
\dot{x}_5 &= -12.147 + 4.049(x_4 - x_2) \\
\dot{x}_6 &= u_1 \\
\dot{x}_7 &= (u_2 - x_7)/2
\end{aligned} \tag{5.60}$$

where the seven elements of the state vector $\mathbf{x}$ are as follows.

x_1 = increment of forward speed (u_A, ft/sec)

x_2 = increment of altitude angle (α, deg)

x_3 = increment of pitch rate ($\dot{\theta}$, deg/sec)

x_4 = increment of pitch angle (θ, deg)

x_5 = height (h, ft)

x_6 = increment of elevator angle (η, deg)

x_7 = increment of throttle (Th)

u_1 = increment of elevator rate ($\dot{\eta}$, deg/sec)

u_2 = input to throttle actuator

and where R the nonlinear ground effect term is

$$R = 400/(3x_5 + 100) - 1$$

In this notation the performance index becomes

$$J = A(x_4 - 8)^2 + \int_0^{20} \{A_1(x_1 - u_D)^2 + A_2(x_5 - h_D)^2 + g_1u_1^2 + g_2(u_2 - 0.75)^{2m}\}\, dt \tag{5.61}$$

where the weighting matrices A, A_1, A_2, g_1, g_2, and m remain to be determined.

The matrices required to implement the gradient algorithm are

$$f_x = \begin{bmatrix} -0.041 & 0.381 & 0.000 & -0.562 & 0.221{*}R_x & 0.000 & -2.552 \\ -0.066 & -0.577 & 1.000 & 0.000 & -0.395{*}R_x & -0.050 & 0.992 \\ 0.011 & -1.108 & -0.822 & 0.000 & -3.607{*}R_x & -1.264 & -0.157 \\ 0.000 & 0.000 & 1.000 & 0.000 & 0.000 & 0.000 & 0.000 \\ 0.000 & -4.049 & 0.000 & 4.049 & 0.000 & 0.000 & 0.000 \\ 0.000 & 0.000 & 0.000 & 0.000 & 0.000 & 0.000 & 0.000 \\ 0.000 & 0.000 & 0.000 & 0.000 & 0.000 & 0.000 & -0.333 \end{bmatrix}$$

where $R_x = 1200/(3x_5 + 100)^2$ and

$$f_u = \begin{bmatrix} 0 & 0 \\ 0 & 0 \\ 0 & 0 \\ 0 & 0 \\ 0 & 0 \\ 1 & 0 \\ 0 & \frac{1}{3} \end{bmatrix}, \qquad L_x = \begin{bmatrix} 2A_1 \ (x_1 - u_D) \\ 0 \\ 1 \\ 2A_2 \ (x_5 - h_D) \\ 0 \\ 0 \\ 0 \end{bmatrix}$$

and

$$L_u = \begin{bmatrix} 2g_1 & u_1 \\ 2g_2 & u_2 + c \end{bmatrix}$$

The weighting values were then determined as follows.

1. Nominal values were chosen and the optimal solution was computed using the gradient method as discussed in Section 5.1 and Examples 5.1 and 5.2.

2. The optimal solution was examined in the light of the requirements just outlined. Then the weighting parameters were adjusted as necessary, e.g., increasing A_2 ensured that the desired height trajectory was followed more closely.

After a few test runs the parameters were chosen to be

$$\begin{aligned} A &= 1; \quad & A_1 &= 1; \quad & A_2 &= 1 \\ g_1 &= \tfrac{1}{3}; \quad & g_2 &= 20; \quad & m &= 1 \end{aligned} \tag{5.62}$$

The convergence of the gradient procedure is shown in Table XX

TABLE XX

CONVERGENCE OF THE GRADIENT METHOD

Iteration Number	Cost	Step-Length Parameter
Priming	224256	
1	7388	0.92×10^{-12}
2	649	0.182×10^{-8}
3	598	0.10×10^{-8}
5	590	0.19×10^{-9}
10	579.8	0.14×10^{-7}
15	579.4	0.14×10^{-11}

together with the value of the step-length parameter used in the algorithm (see also Figure 5.5). Note that the step length changes over several orders of magnitude, emphasizing the need for some automatic step-choice procedure. The height and speed trajectories are shown in

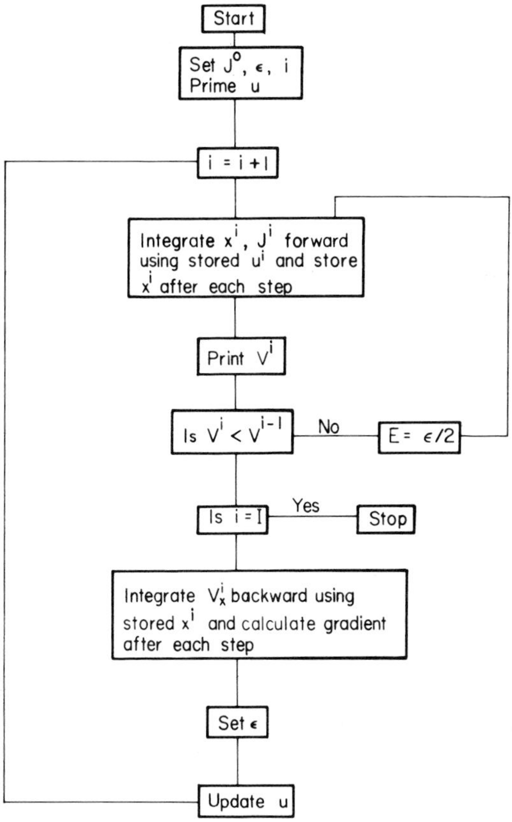

FIG. 5.5. Flow diagram for gradient procedure.

Figure 5.6. Although the performance specifications have been met, the final speed is rather low. In fact it seems likely that the solution is still some way from the optimal. As mentioned previously, this slow final convergence is a characteristic of the gradient procedure.

Numerical integration of the system equations was done with the standard fourth-order Runge–Kutta routine. An integration step length 0.04 sec was used, and one iteration took about 10 sec of IBM 7090 computer time.

Before going on to the next example, the reader is reminded that the solution just obtained is *open loop*. The trajectory would, therefore, be extremely sensitive to perturbations due to wind gusts, etc. A more practical approach would be to linearize the system equations about this trajectory and then to use the linear-quadratic algorithm discussed in Chapter 4. This would give a linear feedback law that could be used in the neighborhood of the optimal. However, as will be seen in the next

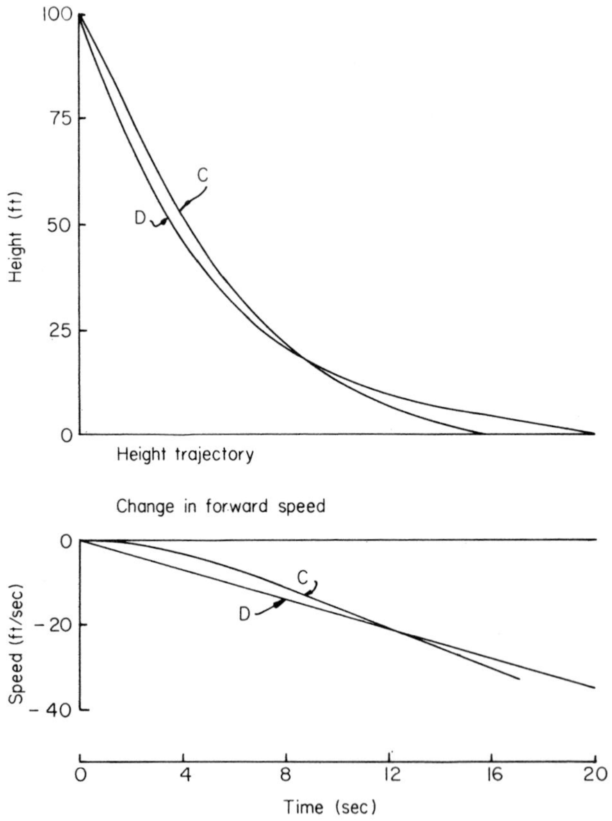

FIG. 5.6. Optimal landing trajectories. *D* is the ideal trajectory, *C* the optimal trajectory.

chapter, a similar solution can be obtained as a product of the computational algorithm used to solve the control problem.

EXAMPLE 5.5. *The Optimal Control of a Boiler.* This example, the optimal control of a steam boiler for an electric power generating system, is also a complex problem even in the somewhat simplified form discussed here. It has, however, the potential of being a rewarding area for study, as relatively small increases in efficiency produce a high economic return.

The problem is also interesting from a pedagogical standpoint because it introduces one of the main drawbacks to the implementation of optimal control: *the lack of an accurate mathematical model.* The reader will recall that even with the airplane examples (4.3 and 5.4) it was tacitly assumed that the mathematical model used was correct. Unfortunately, for many problems, especially those outside the aerospace field, this is not the case.

The choice of a model for the steam boiler will not be discussed here, although some idea of the complexity and the difficulties involved may be obtained from the following description of a typical boiler plant.

The main feed pump drives water through the economizer into the steam separating drum (A) (see Figure 5.7). There it mixes with water at

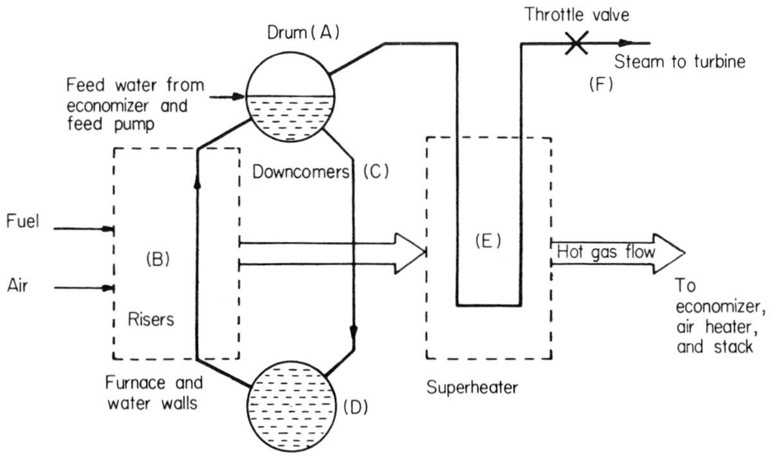

FIG. 5.7. Diagram of the boiler system.

saturation temperature coming from the riser tubes (B) before flowing down the unheated downcomer tubes (C) to the mud drum (D) and back into the risers (B). In modern power plants, the majority of the riser tubes, known as water-walls, surround the furnace walls receiving heat by direct radiation from the burning fuel. As the water passes through the risers it is heated to saturation temperature and a fraction is converted into steam. This separates out in the upper drum (A) and flows through the super-heater (E) to the throttle valve (F), which regulates the steam supply to the turbine. Circulation around the downcomer-riser loop is maintained by the difference in density between the column of steam-water mixture in the riser and the column of water in the downcomer.

The effect of a disturbance, say an increase, in electrical load on the alternator is to reduce the speed of the turbine and alternator. Hence, the frequency of the supply falls. This must be corrected by increasing the steam flow into the turbine and, at present, this is performed automatically by a governor opening the throttle valve. As a result, the boiler pressure falls and the overall energy balance has to be restored by increasing the fuel flow into the furnace. Also, the water level in the separating drum (A) must be maintained by increasing the feedwater flow, although this is complicated by the water level "swelling," due to increased evaporation with reduced pressure, before falling uniformly.

Clearly a description of this plant involves many modes of heat transfer and fluid flow, and some of these physical processes are not amenable to mathematical formulation. However, by using lumped parameters to avoid the difficulties of partial differential equations, as well as certain empirical relationships and other simplifying assumptions, it is possible to account for the principle dynamics with a fifth-order model. Basically the model used here was derived from that presented by Chien *et al.* (1964). The equations are

$$\begin{aligned}
\dot{x}_1 &= (Q_r - (x_1 - x_3)\,w)(66180X/x_2 + (1 - X)/49)/231 \\
\dot{x}_2 &= 427(Xw - x_5) \\
\dot{x}_3 &= (220X + (1 - X)(357.5 + 0.00133x_2) - x_3)\,w/7600 \qquad (5.63) \\
\dot{x}_4 &= (20340x_5/(1 + 0.5x_5) - Q_r)/5700 \\
\dot{x}_5 &= (1.5 + 1.5(2/\pi)\arctan(u) - x_5)110
\end{aligned}$$

where x_1 is the riser outlet enthalpy, x_2 the drum pressure, x_3 is equal to drum liquid enthalpy, x_4 represents the near tube temperature, and x_5 the fuel flow rate, and where

$$\begin{aligned}
T_v &= 388.35 + 0.0014x_2 \\
Q_r &= 0.0722(x_4 - T_v)^3 \\
X &= (s_1 - 357.5 - 0.0133x_2)/715 \qquad (5.64) \\
\rho_r &= (66180X/x_2 + (1 - X)/49)^{-1} \\
w &= [(1176 - 24\rho_r)/(1.947 * 10^{-4} + 2.146 * 10^{-3}/\rho_r)]^{1/2}
\end{aligned}$$

and where u is the control input and the arctan function acts as a limiting function.

The control problem considered is that of minimizing variations in the drum pressure $\dot{x}_2$ and the fuel rate x_5 by controlling the input to the actuator u. A suitable performance criterion is

$$J = \int_0^{150} \{(x_2 - 97920)^2 + 4 \times 10^6 x_5 + 10^4 u^2\}\, dt \tag{5.65}$$

The gradient method may now be used to obtain the optimal control. The partial derivatives needed to compute the derivative of the return V_x may be obtained by differentiating Equations (5.63) and (5.65).

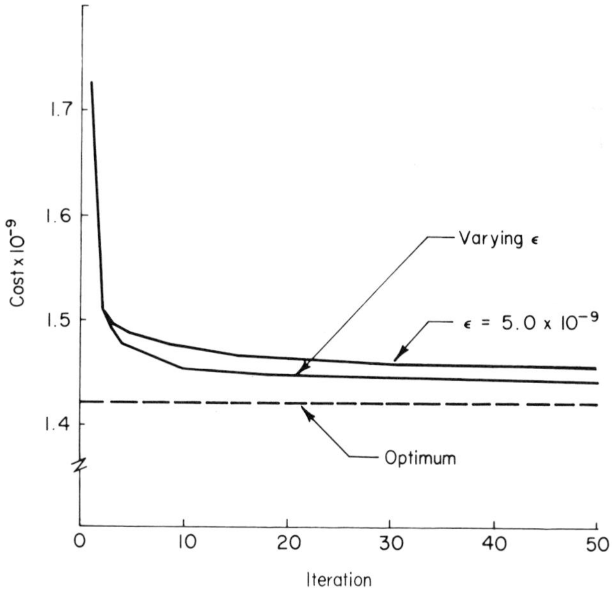

FIG. 5.8. Convergence of the gradient algorithm.

Several runs were made to illustrate some of the characteristics of the gradient procedure. For example, two runs were made to show how the choice of the step length affects the convergence. In one run the step length was a constant and in the second it was chosen automatically (increasing after a successful iteration and decreasing after an unsuccessful iteration). The convergence is shown in Figure 5.8. As can be seen, the difference is marked, but even with a varying step length the optimum was not reached in 50 iterations. Several attempts were made to speed up the convergence but without much success.

Some near optimal trajectories are shown in Figure 5.9, and further results may be found in the work by Markland (1966).

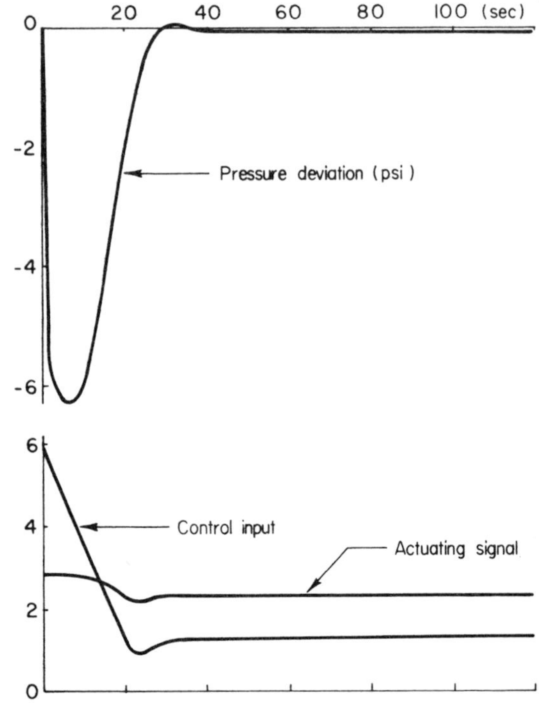

FIG. 5.9. Optimal control. (Step change in steam load, 90 to 100 percent full load.)

Numerical integration of the system and characteristic equations was again done with a standard Runge–Kutta routine. An integration step length of 0.5 sec was found to give sufficient accuracy so that the 300 integration steps were required. Linear interpolation was used in the backward integration to improve the accuracy. With the foregoing integration step length, one iteration took approximately 14 sec on an IBM 7040 computer.

5.4 THE FIRST VARIATION: A STATIONARITY CONDITION FOR A LOCAL OPTIMUM

It is fairly clear that the gradient algorithm will stop iterating when the term $V_x f_u + L_u$ is identically zero along the nominal trajectory. This condition

$$V_x f_u + L_u = 0 \tag{5.66}$$

is the *stationarity* condition for a local optimum, which will now be derived from a consideration of the first variation of the performance index. The derivation is very similar to the derivation of the gradient method that was just given.

Let $u^0(t)$ denote some nominal control and $V^0(x, t)$ be the related return function, i.e.,

$$V^0(x_0, t_0) = \int_{t_0}^{t_f} L(x, u^0)\, dt + \phi(x(t_f), t_f) \tag{5.67}$$

with $x(t_0) = x_0$ specified and

$$\dot{x} = f(x, u^0, t) \tag{5.68}$$

Let $u^1(t)$ denote a control in the neighborhood of $u^0(t)$, i.e., $u^1(t) = u^0(t) + \epsilon\eta(t)$ where the positive parameter ϵ is chosen such that $| u^1 - u^0 | < \delta$, where δ may be arbitrarily small.

Now by employing arguments similar to those in Section 5.2, with $t_1 = t_0$ and $t_m = t_f$, the difference between the cost obtained by using the control u^1 and the cost obtained by using the control u^0 may be shown to be given by

$$J(u^1) - J(u^0) = \delta J = \int_{t_0}^{t_f} \{V_t^0(x^1, t) + V_x^0(x^1, t) f(x^1, u^1, t) + L(x^1, u^1, t)\}\, dt \tag{5.69†}$$

where V^0 is the return function defined by the nominal control, i.e.,

$$\begin{aligned} V_t^0(x, t) + V_x^0(x, t) f(x, u^0, t) + L(x, u^0, t) &= 0 \\ V^0(t_f) &= \phi(t_f) \end{aligned} \tag{5.70}$$

and where $x^1(t)$ is the trajectory determined by the new control $u^1(t)$. Now, using this partial differential equation, Equation (5.70), to eliminate $V_t^0(x^1, t)$ from Equation (5.69), and then expanding Equation (5.69) around the nominal control and trajectory, we obtain

$$J = \int_{t_f}^{t^0} [V_x^0(x^0, t) f_u(x^0, u^0, t) + L_u(x^0, u^0, t)]\, \delta u + 0(\delta u^2, \delta u\, \delta x, \delta x^2) \tag{5.71}$$

Substituting for $\delta u = u^1 - u^0 = \epsilon\eta$, we obtain

$$\delta J = \epsilon \int_{t_0}^{t_f} \{V_x^0 f_u + L_u\}\, \eta\, dt + O(\epsilon^2)$$

† The reader should note the difference between Equation (5.69) and Equation (5.15). Here the return function V^0 is defined essentially by the *nominal* control. In Equation (5.15) the return was defined by the *new nominal* control. The two expressions are, of course, identical. It is just that here V^0 is a more convenient return function to use. It is left to the reader to complete the arguments to justify Equation (5.69).

The first term on the right-hand side of this expression is referred to as the first variation, $\delta^1 J$, i.e.,

$$\delta^1 J = \epsilon \int_{t_0}^{t_f} \{V_x{}^0 f_u + L_u\} \, \eta \, dt$$

Clearly, for u^0 to be an extremal solution, $\delta^1 J$ must be equal to zero (cf Section 2.2). But for $\delta^1 J$ to be zero for all variations, η, the following condition must hold

$$V_x{}^0 f_u + L_u = 0 \qquad \text{for all} \quad t[t_0, t_f] \tag{5.72}$$

As noted before, this is the *stationarity condition* which must hold for a local extremal solution provided no constraints are placed on the variations δu.

An equation for $V_x{}^0(t)$ may be derived by partially differentiating Equation (5.70) with respect to x. Thus, V_x is given by

$$\begin{aligned} \dot{V}_x{}^0 &= V_{tx}^0(x^0, t) + V_{xx}^0(x^0, t) f(x^0, u^0, t) \\ &= -V_x{}^0(x^0, t) f_x(x^0, u^0, t) - L_x(x^0, u^0, t) \end{aligned} \tag{5.73}$$

with $V_x{}^0(x^0, t_f) = \phi_x(x^0, t_f)$.

This is exactly the same set of equations as was used to compute V_x in the gradient algorithm [cf Equations (5.10) and (5.11)].

Next, a classical derivation of the first variation will be presented. The performance index is augmented by adjoining the system equations as follows

$$J^* = \phi(x(t_f), t_f) + \int_{t_0}^{t_f} \{L(x(t), u(t), t) + p^{\mathrm{T}}(t)(f(x, u, t) - \dot{x})\} \, dt \tag{5.74}$$

where

$$p(t) = \begin{bmatrix} p_1(t) \\ \vdots \\ p_n(t) \end{bmatrix}$$

is an n-dimensional vector, the *adjoint* or *costate* vector, which is to be specified later. For convenience, the Hamiltonian notation will be used, viz,

$$H(x, p, u, t) = p(t)^{\mathrm{T}} f(x, u, t) + L(x, u, t) \tag{5.75}$$

where H is the so-called Hamiltonian. Thus, Equation (5.73) may be written $J^* = \Phi(x(t_f)) + \int_{t_0}^{t_f} \{H - p^T\dot{x}\}\, dt$. Now integrating $-p^T\dot{x}$ by parts leads to

$$J^* = \int_{t_0}^{t_f} \{H + \dot{p}^T x\}\, dt + \Phi(x(t_f), t_f) + p(t_0)^T x(t_0) - p(t_f)^T x(t_f)$$

Next consider a first-order perturbation in J^* about some nominal control u^0 and state x^0, i.e.,

$$J^*(x, u) = J^*(x^0, u^0) + [\phi_x - p(t_f)^T]\, \delta x\,(t_f)$$
$$+ p(t_0)^T x(t_0) + \int_{t_0}^{t_f} \{[H_x + \dot{p}]^T \delta x + H_u\, \delta u\}\, dt + O(\epsilon^2) \qquad (5.76)$$

Now choosing $p(t)$ such that $p(t_f) = \phi_x$, $\dot{p} = -H_x$, and noting that $\delta x(t_0) = 0$ leads to

$$\delta J^* = J(x, u) - J(x^0, u^0) = \int_{t_0}^{t_f} \{H_u\, \delta u\}\, dt \qquad (5.77)$$

Thus the first variation, $\delta^1 J$, is

$$\delta^1 J = \int_{t_0}^{t_f} \{H_u\, \delta u\}\, dt \qquad (5.78)$$

Clearly Equations (5.78) and (5.71) are identical since the costate vector $p(t)$ can be identified with the partial derivative $V_x(x^0(t), t)$.†

If side constraints are present the stationarity condition becomes

$$\int_{t_0}^{t_f} (V_x f_u + L_u)\, \delta u^*\, dt = 0 \qquad (5.79)$$

where the δu^* are *admissible* variations in u. The variations δu^* must be such that the side constraint is unchanged, at least to first order. As the reader can easily verify [cf Equation (5.36)] this implies that δu^* must be chosen so that

$$\int_{t_0}^{t_f} (W_x f_u + M_u)\, \delta u^* = 0 \qquad (5.80)$$

Formally adjoining this constraint to Equation (5.79) with a Lagrange multiplier ν gives

$$\int_{t_0}^{t_f} \{V_x f_u + L_u + \nu^T(W_x f_u + M_u)\}\, \delta u\, dt = 0$$

† It is left to the reader to verify that $p(t) = V_x(x, t)$.

where δu may now be considered to be an arbitrary function. Thus the stationarity condition may be written

$$V_x f_u + L_u + \nu^{\mathrm{T}}(W_x f_u + M_u) = 0 \tag{5.81}$$

Finally, the first variation at the terminal time must be examined. This variation $\delta^1 V(t_f)$ may be shown to be [cf Equation (5.43)]

$$\begin{aligned}\delta^1 V(t_f) = [&L(x(t_f), u(t_f), t_f) + \Phi_t(x(t_f), \nu, t_f) \\ &+ \Phi_x(x(t_f), \nu, t_f) f(x(t_f), u(t_f), t_f)]\, dt_f\end{aligned} \tag{5.82}$$

Thus the stationarity condition for the end time t_f is

$$0 = L(x(t_f), u(t_f), t) + \Phi_t(x(t_f), \nu, t_f) + \Phi_x(x(t_f), \nu, t_f) f(x(t_f), u(t_f), t_f)$$

PROBLEMS

1. Using the finite difference technique, derive the gradient algorithm for systems with control variables *and* control parameters.
2. Program the gradient algorithm and use the program to duplicate the results of Examples 1 and 2.
3. Using the program developed in Problem 2, examine the various gradient correction algorithms e.g.,

(a) $\delta u = \epsilon \operatorname{sgn}(V_x f_u + L_u)$

(b) $\delta u = \epsilon(V_x f_u + L_u)$

(c) $\delta u = \epsilon(V_x f_{uu} + L_{uu})^{-1}(V_x f_u + L_u)$

4. Using the classical variation approach, derive the stationarity condition for a local optimum for the problem with side constraints.

BIBLIOGRAPHY AND COMMENTS

The gradient method for dynamic optimization problems was first described in

Bryson, A. E. and Denham, W. F. (1962). "A Steepest Ascent Method for Solving Optimum Programming Problems," *J. Appl. Mech.*, Vol. 29, pp. 247–257.

Kelley, H. J. (1962). "Method of Gradients," *Optimization Techniques* (G. Leitmann, ed.), Chap. 6. Academic Press, New York.

Section 5.3. The aircraft landing problem and the boiler problem were described in

Noton, A. R. M., Dyer, P., and Markland, C. A. (1966). "The Numerical Computation of Optimal Control," *Proc. Joint Autom. Control Conf.*, Seattle, Washington; and *Trans. IEEE Autom. Control*, pp. 193–204. February, 1967.

There are many other examples of the gradient method in the literature. See, for example,

Balakrishnan, A. V. and Neustadt, L. W. (1964). *Computing Methods in Optimization Problems*. Academic Press, New York.

Merriam, C. W. (1964). *Optimization Theory and the Design of Feedback Control Systems*. McGraw-Hill, New York.

Markland, C. A. (1966). "Optimal Control of a Boiler," Ph.D. thesis. Univ. of Nottingham, England.

The boiler model used here was given by

Chien, K. L., Ergin, E. I., Ling, C., and Lee, A. (1958). "Dynamic Analysis of a Boiler," *Trans. A.S.M.E., J. of Basic Eng.*, Vol. 80, pp. 1809–1819.

6

The Successive Sweep Method and the Second Variation

6.1 INTRODUCTION

In this chapter a Newton–Raphson algorithm is developed for continuous optimal control problems. The algorithm is referred to as the successive sweep algorithm and is derived from the continuous dynamic programming equations. The analysis is similar to that employed in deriving the gradient algorithm except that the expansions are carried out to second order. The algorithm is developed for problems with side constraints, control parameters, and free terminal time. It is applied to the examples considered in the last chapter: the brachistochrone problem, an aircraft landing problem, and the optimal control of a boiler. This enables a comparison of the gradient method and the successive sweep method to be made.

The technique of employing second-order expansions leads to a novel developement of the second variation. Necessary and sufficient conditions for the negative definiteness of the second variation are obtained, thus completing the theory concerning local optimality of optimal control problems.

6.2 THE SUCCESSIVE SWEEP METHOD

The successive sweep method is motivated, as was the gradient method, by a consideration of an expansion of the return function about some nominal solution. The algorithm can, of course, be obtained directly from the discrete version using the finite difference approach, although the algebra is somewhat tedious. However, another approach is to employ continuous dynamic programming directly.

Following the development of the gradient method, it is assumed that a new nominal control function $u^{j+1}(t)$ has been found for times greater than t_m. The return function associated with the control u^{j+1}, is, at the time t_m, denoted by $V(x, t_m)$. The problem is to extend this new control u^{j+1} to the left of t_m while increasing the value of the return.

Suppose that at some time t_1 prior to t_m, the nominal control u^j is replaced by the new control u^{j+1} (the control u^{j+1} is used for *all* times after t_m) (Figure 6.1). It was shown in Section 5.2, Equation (5.17), that

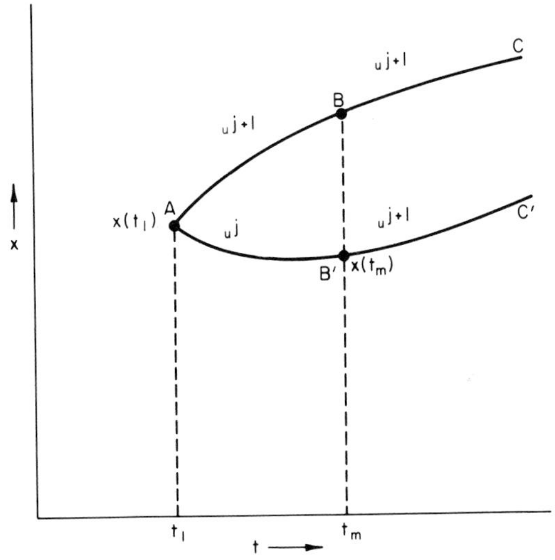

FIG. 6.1. The successive sweep method.

the difference in the return δV due to the change in control from u^j to u^{j+1} is given by

$$\delta V(t_1) = \int_{t_1}^{t_m} \{V_x(\hat{x}, t)[f(\hat{x}, u^{j+1}, t) - f(\hat{x}, u^j, t)] + L(\hat{x}, u^{j+1}, t) - L(\hat{x}, u^j, t)\}\, dt \tag{6.1}$$

where the integral is integrated along the path $\hat{x}$ determined by

$$\dot{\hat{x}} = f(\hat{x}, u^j, t), \qquad \hat{x}(t_1) = x^{j+1}(t_1) \tag{6.2}$$

and where the return function V is defined essentially by u^{j+1}, i.e.,

$$\begin{aligned} V(x, t_f) &= \phi(x, t_f) \\ V_t(x, t) + V_x(x, t) f(x, u^{j+1}, t) + L(x, u^{j+1}, t) &= 0 \end{aligned} \tag{6.3}$$

Now the right-hand side of Equation (6.1) is expanded about the nominal control u^j and the nominal trajectory x^j to second order in δx and δu [cf Equation (5.27)] giving

$$\delta V(t_1) = \int_{t_1}^{t_m} \left\{ Z_u \, \delta u + \tfrac{1}{2}[\delta x^{\mathrm{T}} \, \delta u^{\mathrm{T}}] \begin{bmatrix} 0 & Z_{xu} \\ Z_{ux} & Z_{uu} \end{bmatrix} \begin{bmatrix} \delta x \\ \delta u \end{bmatrix} \right\} dt \tag{6.4}$$

where $\delta x = x^{j+1} - x^j$ and $\delta u = u^{j+1} - u^j$. The m-vector Z_u, the $n \times m$ matrix Z_{xu}, and the $m \times m$ matrix Z_{uu} are given by†

$$\begin{aligned} Z_u &= V_x f_u + L_u \\ Z_{xu} &= V_{xx} f_u + V_x f_{xu} + L_{xu} \\ Z_{uu} &= V_x f_{uu} + L_{uu} \end{aligned} \tag{6.5}$$

where the partial derivatives L_u, V_x, etc., are evaluated along the nominal trajectory x^j, u^j. Letting t_1 approach t_m so that $dt = t_m - t_1$ becomes small allows the integral equation (6.4) to be approximated to first order by

$$\delta V(t_1) = \left\{ Z_u \delta u(t_1) + \tfrac{1}{2}[\delta x(t_1)^{\mathrm{T}} \, \delta u(t_1)^{\mathrm{T}}] \begin{bmatrix} 0 & Z_{xu} \\ Z_{ux} & Z_{uu} \end{bmatrix} \begin{bmatrix} \delta x(t_1) \\ \delta u(t_1) \end{bmatrix} \right\} dt \tag{6.6}$$

Clearly, $\delta V(t_1)$ will be as large as possible if $\delta u(t_1)$ is chosen to maximize the right-hand side of Equation (6.6), i.e.,

$$\delta u(t_1) = -Z_{uu}^{-1}[Z_u + Z_{ux} \, \delta x(t_1)]$$

and, more generally,

$$\delta u(t) = -Z_{uu}^{-1}[Z_u + Z_{ux} \, \delta x(t)], \qquad t_0 \leqslant t \leqslant t_f \tag{6.7}$$

As can be seen, a basic difference between this correction and the gradient correction, Equation (5.26), is the additional term $-Z_{uu}^{-1} Z_{ux} \, \delta x$ in the sweep algorithm. Note that this means that the time-varying $m \times n$ matrix $-Z_{uu}^{-1} Z_{ux}$ must be stored along the trajectory. Also the differential equation for $V_x(t)$ is changed because δu depends on δx.

† Note that the subscripts on the function Z are included for notational convenience and are not intended to indicate partial derivatives. Note however that the functions Z_u, Z_{xu}, and Z_{uu} may be considered as partial derivatives of the function $Z = V_x f + L$.

It remains to find the differential equations for the partial derivatives $V_x(x^j, t)$ and $V_{xx}(x^j, t)$. The terminal conditions are obtained by differentiating the boundary condition $V(x, t_f) = \phi(x, t_f)$, viz,

$$V_x(x^j, t_f) = \phi_x(x^j, t_f), \qquad V_{xx}(x^j, t_f) = \phi_{xx}(x^j, t_f) \tag{6.8}$$

For times prior to the final time t_f, the differential equations for V_x and V_{xx} may be derived from the partial differential equation for the return, Equation (6.3).[†] First, this partial differential equation is expanded to second order about the nominal control and state u^j and x^j

$$\begin{aligned}0 = V_t + V_x f + L + [V_{tx} + V_{xx} f + V_x f_x + L_x]\,\delta x \\ + [V_x f_u + L_u]\,\delta u + \tfrac{1}{2}\delta x^{\mathrm{T}}[V_{txx} + V_{xxx} f + V_{xx} f_x + f_x{}^{\mathrm{T}} V_{xx} \\ + L_{xx} + V_x f_{xx}]\,\delta x + \delta x^{\mathrm{T}}[V_{xx} f_u + V_x f_{xu} + L_{xu}]\,\delta u \\ + \tfrac{1}{2}\delta u^{\mathrm{T}}[V_x f_{uu} + L_{uu}]\,\delta u \cdots\end{aligned} \tag{6.9}$$

where all the partial derivatives are evaluated along the nominal. This expression may be simplified by noting that the time derivatives of V_x and V_{xx}, along x^j, may be expressed as

$$\dot{V}_x = V_{tx} + V_{xx} f \quad \text{and} \quad \dot{V}_{xx} = V_{txx} + V_{xxx} f \tag{6.10}$$

It is also convenient to define the functions Z_x and Z_{xx}

$$\begin{aligned}Z_x &= V_x f_x + L_x \\ Z_{xx} &= V_x f_{xx} + V_{xx} f_x + f_x{}^{\mathrm{T}} V_{xx} + L_{xx}\end{aligned} \tag{6.11}$$

Substituting Equations (6.10) and (6.11), Equation (6.9) can be written

$$\begin{aligned}0 = [V_t + V_x f + L] + [\dot{V}_x + Z_x]\,\delta x \\ + Z_u\,\delta u + \tfrac{1}{2}\delta x^{\mathrm{T}}[\dot{V}_{xx} + Z_{xx}]\,\delta x \cdots \\ + \delta x^{\mathrm{T}} Z_{xu}\,\delta u + \tfrac{1}{2}\delta u^{\mathrm{T}} Z_{uu}\,\delta u\end{aligned}$$

Now using Equation (6.7) to eliminate δu, one obtains

$$\begin{aligned}0 = [V_t + V_x f + L] + [\dot{V}_x + Z_x - Z_u Z_{uu}^{-1} Z_{ux}]\,\delta x \\ + \tfrac{1}{2}\delta x^{\mathrm{T}}[\dot{V}_{xx} + Z_{xx} - Z_{xu} Z_{uu}^{-1} Z_{ux}]\,\delta x \cdots\end{aligned} \tag{6.12}$$

Since this expression may be considered as an identity, each term in

[†] This technique is usually referred to as the *method of characteristics*.

brackets may be set to zero,[†] which gives the following differential equations for V_x and V_{xx}

$$\dot{V}_x = -Z_x + Z_u Z_{uu}^{-1} Z_{ux}, \qquad \dot{V}_{xx} = -Z_{xx} + Z_{xu} Z_{uu}^{-1} Z_{ux} \tag{6.13}$$

The basic successive sweep algorithm may be summarized as follows.

1. Choose some nominal control u^j and integrate the state equations forward storing the state trajectory x^j. Compute the cost J^j.

2. Using the stored values of the state x^j and control u^j, compute V_x and V_{xx} *backward* in time using the Equations (6.13). At the same time, compute and store the control correction δu^0, $\delta u^0 = -Z_{uu}^{-1} Z_u{}^{\mathrm{T}}$ and the feedback matrix u_x, $u_x = -Z_{uu}^{-1} Z_{ux}$.

3. Integrate the state equations forward using the new control u^{j+1} where

$$u^{j+1} = u^j + \epsilon\, \delta u^0 + u_x(x^{j+1} - x^j) \tag{6.14}$$

where ϵ is some positive constant in the range (0,1). Compute the new cost J^{j+1}.

4. If $J^{j+1} > J^j$, repeat step 2 with $j = j + 1$. If $J^j > J^{j+1}$, reduce the step size ϵ and repeat step 3.

5. Steps 2 and 3 are repeated until the optimal solution is reached.

It should be noted that the step-size parameter ϵ is *not equivalent* here to the step size in the gradient method. For, in the vicinity of the optimal solution, $\epsilon = 1$ will ensure convergence. With the gradient method, even if the nominal is near to the optimal, the step size is unknown.

It should also be noted that if the system equations are linear and the performance criteria is quadratic, *one step convergence* can be achieved with $\epsilon = 1$.

6.3 AN ALTERNATIVE DERIVATION: CONTROL PARAMETERS

The successive sweep algorithm can also be obtained directly from the dynamic programming equations for the optimal return function and the optimal control law. This approach will now be outlined, and dependence on control parameters α will be included.

Let $V(x, \alpha, t)$ be the nominal return function at a given time t. The best control to employ is the one that maximizes $V_x(x, \alpha, t)$

[†] Not all of the $\delta x^{\mathrm{T}}\, \delta x$ terms are included. However, the missing terms are multiplied by Z_u, which is considered to be small. Thus these terms may be omitted without affecting the convergence of the algorithm.

$f(x, u, \alpha, t) + L(x, u, \alpha, t)$ [cf Equation 4.14]. The successive sweep method approximates this control by choosing a correction to a nominal control in order to maximize a second-order expansion of this function. Thus, this correction, denoted by δu^*, is given by

$$\delta u^{\text{opt}} = \arg\max_{\delta u}\{Z_u\,\delta u + Z_\alpha\,\delta\alpha + (V_{xx}f + Z_x)\,\delta x + \tfrac{1}{2}[\delta x^{\mathrm{T}}\ \delta\alpha^{\mathrm{T}}\ \delta u^{\mathrm{T}}]\begin{bmatrix} Z_{xx} + V_{xxx}f & \text{symmetric} & \\ Z_{x\alpha} + V_x f_x & Z_{\alpha\alpha} + V_{\alpha\alpha x}f & \\ Z_{xu} & Z_{\alpha u} & Z_{uu}\end{bmatrix}\begin{bmatrix}\delta x\\ \delta\alpha\\ \delta u\end{bmatrix} \tag{6.15}$$

where Z_u, Z_x, Z_{uu}, Z_{xx}, and Z_{ux} are given by Equations (6.5) and (6.10) and Z_α, $Z_{\alpha\alpha}$, $Z_{\alpha x}$, $Z_{\alpha u}$ are

$$\begin{aligned} Z_\alpha &= V_x f_\alpha + L_\alpha \\ Z_{x\alpha} &= f_x{}^{\mathrm{T}} V_{x\alpha} + V_x f_{x\alpha} + V_{xx} f_\alpha + L_{x\alpha} \\ Z_{\alpha u} &= V_{\alpha x} f_u + L_{\alpha u} + V_x f_{\alpha u} \\ Z_{\alpha\alpha} &= V_{\alpha x} f_\alpha + f_\alpha{}^{\mathrm{T}} V_{x\alpha} + V_x f_{\alpha\alpha} + L_{\alpha\alpha} \end{aligned} \tag{6.16}$$

Hence, δu^{opt} is given by

$$\delta u^{\text{opt}} = -Z_{uu}^{-1}(Z_u + Z_{ux}\,\delta x + Z_{u\alpha}\,\delta\alpha) \tag{6.17}$$

Note, as might be expected from the analysis of the previous section, that the inclusion of the control parameters is reflected in the correction policy δu^{opt}.

Characteristic equations for the partial derivatives V_x, V_α, V_{xx}, etc., could now be derived as before. However, a different approach will be taken here. For one of the disadvantages of this algorithm is that the time-varying feedback gains $Z_{uu}^{-1}Z_u$, $Z_{uu}^{-1}Z_{ux}$, and $Z_{uu}^{-1}Z_{u\alpha}$ must be stored along the whole trajectory. Obviously, the storage requirements for a high-order system may be excessive. One approach to this subsiduary problem, used by the authors, is to use *constant* feedback gains. Thus, each time-varying feedback gain is replaced by a set of one or more constants. Clearly this approach is suboptimal; however, in practice little is lost in terms of speed of convergence.

Of course, the characteristic equations are modified by this approach. For notational convenience the expression for δu^{opt} is written

$$\delta u^{\text{opt}}(x, \alpha, t) = \delta u^0 + u_x\,\delta x + u_\alpha\,\delta\alpha \tag{6.18}$$

where optimal values of δu, u_x, and u_y are

$$\delta u^0 = -Z_{uu}^{-1}Z_u, \qquad u_x = -Z_{uu}^{-1}Z_{ux}, \qquad u_\alpha = -Z_{uu}^{-1}Z_{u\alpha} \tag{6.19}$$

Now $V(x, \alpha, t)$ satisfies the boundary value problem

$$V(x(t_f), \alpha, t_f) = \phi(x(t_f), \alpha, t_f)$$
$$V_t + V_x f(x, u(x, \alpha, t), \alpha, t) + L(x, u(x, \alpha, t), \alpha, t) = 0 \tag{6.20}$$

where

$$u = u^j + \delta u^{\mathrm{opt}} \tag{6.21}$$

Expanding the foregoing equations about the nominal, we obtain the characteristic equations for V_x, V_α, etc., viz,

$$\begin{aligned}
\dot{V}_x &= -Z_x - Z_u u_x - [Z_{xu}\,\delta u^0]^{\mathrm{T}} - \delta u^0 Z_{uu} u_x, & V_x(t_f) &= \phi_x(t_f)\\
\dot{V}_\alpha &= -Z_\alpha - [Z_{\alpha u}\,\delta u^0]^{\mathrm{T}} - Z_u u_\alpha - \delta u^0 Z_{uu} u_\alpha, & V_\alpha(t_f) &= \phi_\alpha(t_f)\\
\dot{V}_{x\alpha} &= -Z_{x\alpha} - Z_{xu} u_\alpha - u_x^{\mathrm{T}} Z_{u\alpha} - u_x^{\mathrm{T}} Z_{uu} u_\alpha, & V_{x\alpha}(t_f) &= \phi_{x\alpha}(t_f)\\
\dot{V}_{\alpha\alpha} &= -Z_{\alpha\alpha} - Z_{\alpha u} u_\alpha - u_\alpha^{\mathrm{T}} Z_{u\alpha} - u_\alpha^{\mathrm{T}} Z_{uu} u_\alpha, & V_{\alpha\alpha}(t_f) &= \phi_{\alpha\alpha}(t_f)\\
\dot{V}_{xx} &= -Z_{xx} - Z_{xu} u_x - u_x^{\mathrm{T}} Z_{ux} - u_x^{\mathrm{T}} Z_{uu} u_x, & V_{xx}(t_f) &= \phi_{xx}(t_f)
\end{aligned} \tag{6.22}$$

where the Z_α, etc., have just been defined.

The choice of the constant feedback gains u_x, u_α is somewhat arbitrary. One procedure is as follows: At the final time the first set of gains u_x, u_α are chosen optimally [cf Equation (6.19)]. Then, as the backward integration proceeds these gains are compared with $-Z_{uu}^{-1}Z_{ux}$ and $-Z_{uu}^{-1}Z_{u\alpha}$ which are computed, although *not* stored, at each integration step. When a significant (in some sense) difference appears, say at time t_1, the feedback gains are updated using

$$u_x = -Z_{uu}^{-1}(t_1)\, Z_{ux}(t_1), \qquad u_\alpha = -Z_{uu}^{-1}(t_1)\, Z_{u\alpha}(t_1) \tag{6.23}$$

However, δu^0 is chosen optimally and the characteristic equations for V_x and V_α simplify to

$$\dot{V}_x = -Z_x - Z_{xu} Z_{uu}^{-1} Z_u, \qquad \dot{V}_\alpha = -Z_\alpha - Z_{\alpha u} Z_{uu}^{-1} Z_u \tag{6.24}$$

Finally, the correction to α^j, $\delta\alpha^{\mathrm{opt}}$ is obtained by maximizing a second-order expansion of the return at the initial time, i.e.,

$$\delta\alpha^{\mathrm{opt}} = -V_{\alpha\alpha}^{-1}(x(t_0), \alpha, t_0)\, V_\alpha^{\mathrm{T}}(t_0) \tag{6.25}$$

Thus, the inclusion of control parameters necessitates a few modifications to the basic procedure. First, when updating the control the rule,

$$^{j+1} = u^j + \delta u^0 + u_x\,\delta x + u_\alpha\,\delta\alpha \tag{6.26}$$

is used. Secondly, at the initial time the new nominal control parameter α^{j+1} is computed by

$$\alpha^{j+1} = \alpha^j + \delta\alpha^{\text{opt}} \tag{6.27}$$

where $\delta\alpha^{\text{opt}}$ is given in Equation (6.25).

6.3.1 Free Initial Conditions

It was assumed before that the initial state $x(t_0)$ was specified, which may not always be the case. If the initial state is not specified, the corrections to the control parameters and to the initial state $\delta x(t_0)$, $\delta\alpha$ are chosen by

$$\begin{bmatrix} \delta x(t_0) \\ \delta\alpha \end{bmatrix} = -\begin{bmatrix} V_{xx}(t_0) & V_{x\alpha}(t_0) \\ V_{\alpha x}(t_0) & V_{\alpha\alpha}(t_0) \end{bmatrix}^{-1} \begin{bmatrix} V_x^{\mathrm{T}}(t_0) \\ V_\alpha^{\mathrm{T}}(t_0) \end{bmatrix} \tag{6.28}$$

6.3.2 Side Constraints

If side constraints of the form

$$\Psi(x(t_f), \alpha, t_f) = 0 \tag{6.29}$$

are present, they may be adjoined to the performance index with a set of Lagrange multipliers ν to form an augmented index J^* where

$$J^* = \phi(x(t_f), \alpha, t_f) + \nu^{\mathrm{T}}\Psi(x(t_f), \alpha, t_f) + \int_{t_0}^{t_f} L(x, u, \alpha, t)\, dt \tag{6.30}$$

The unknown Lagrange multipliers are then treated in a similar fashion to control parameters. Note, however, that whereas J^* is maximized with respect to the control parameters, it is *extremized* with respect to the multipliers. It can be shown easily that the correction for u^j, δu^{opt}, becomes

$$\delta u^{\text{opt}} = \delta u^0 + u_x\, \delta x + u_\alpha\, \delta\alpha + u_\nu\, \delta\nu \tag{6.31}$$

where δu^0, u_x, and u_α have just been defined and where

$$u_\nu = -Z_{uu}^{-1} Z_{u\nu} \tag{6.32}$$

with the $m \times p$ matrix $Z_{u\nu}$ given by

$$Z_{u\nu} = f_u^{\mathrm{T}} V_{x\nu} \tag{6.33}$$

The characteristic equation for $V_{x\nu}$ may be shown to be

$$\dot{V}_{x\nu} = -Z_{x\nu} - Z_{xu}u_\nu - u_x^{\mathrm{T}}Z_{u\nu} - u_x^{\mathrm{T}}Z_{uu}u_\nu, \qquad V_{x\nu}(t_f) = \Psi_x^{\mathrm{T}} \tag{6.34}$$

where the $n \times p$ matrix $Z_{x\nu}$ is $Z_{x\nu} = f_x^{\mathrm{T}} V_{x\nu}$.

The correction $\delta\nu$ is obtained by extremizing a second-order expansion of the return $V(x, \alpha, \nu, t)$, i.e.,

$$\delta\nu = -V_{\nu\nu}^{-1}(t)(V_{\nu x}(t)\,\delta x(t) + V_{\nu}^{\mathrm{T}}(t)) \tag{6.35}$$

where

$$Z_{\nu\alpha} = V_{x\nu}^{\mathrm{T}} f_{\alpha} \tag{6.36}$$

and the characteristic equations for $V_{\nu\nu}$ and V_{ν} are

$$\begin{aligned} \dot{V}_{\nu\nu} &= -Z_{\nu u}u_{\nu} - u_{\nu}^{\mathrm{T}}Z_{u\nu} - u_{\nu}^{\mathrm{T}}Z_{\nu\nu}u_{\nu}\,, \qquad V_{\nu\nu}(t_f) = 0 \\ \dot{V}_{\nu\alpha} &= -Z_{\nu\alpha} - Z_{\nu u}u_{\alpha} - u_{\nu}^{\mathrm{T}}Z_{u\alpha} - u_{\nu}^{\mathrm{T}}Z_{uu}u_{\alpha}\,, \qquad V_{\nu\alpha}(t_f) = \Psi_{\alpha} \end{aligned} \tag{6.37}$$

$$\dot{V}_{\nu} = -Z_{\nu u}Z_{uu}^{-1}Z_{u}\,, \qquad V_{\nu}(t_f) = \Psi \tag{6.38}$$

Now $\delta\nu$ is a constant for any particular trajectory, and hence the expression for $\delta\nu$, Equation (6.35), may be evaluated at any time along the trajectory. However, as $V_{\nu\nu}(t_f) = 0$, it is clear that caution is required in determining when the $\delta\nu$'s may be obtained. If the theoretical gain $u_{\nu} = -Z_{uu}^{-1}Z_{u\nu}$, Equation (6.32), is used, Equation (6.31) becomes

$$\dot{V}_{\nu\nu} = -Z_{\nu u}Z_{uu}^{-1}Z_{u\nu}\,, \qquad V_{\nu\nu}(t_f) = 0 \tag{6.39}$$

$$\dot{V}_{\nu\alpha} = -Z_{\nu\alpha} - Z_{\nu u}u_{\alpha}\,, \qquad V_{\nu\alpha}(t_f) = \Psi_{\alpha} \tag{6.40}$$

For a maximum, Z_{uu} will be negative so that $V_{\nu\nu}$ will become more positive as the backward integration proceeds. Hence, $V_{\nu\nu}$ has maximum rank at the initial time. If $V_{\nu\nu}$ is singular throughout the trajectory, the problem is abnormal. As was discussed in Chapter 4, abnormality implies that the side constraints are functionally interdependent or that the system is uncontrollable.

From a computational point of view, the presence of side constraints means that (for p constraints) $p + p^*n + p(p+1)/2$ extra equations must be included in the backward integration. Extra storage will also be required. However, as soon as the matrix $V_{\nu\nu}$ can be inverted, the dependence on ν may be eliminated. For the second-order expansion of $V(x, \alpha, \nu, t)$ is

$$\begin{aligned} V &= V^0 + V_x\,\delta x + V_{\alpha}\,\delta\alpha + V_{\nu}\,\delta\nu \\ &\quad + \tfrac{1}{2}[\delta x^{\mathrm{T}}, \delta\alpha^{\mathrm{T}}, \delta\nu^{\mathrm{T}}] \begin{bmatrix} V_{xx} & \text{symmetric} & \\ V_{\alpha x} & V_{\alpha\alpha} & \\ V_{\nu x} & V_{\alpha\nu} & V_{\nu\nu} \end{bmatrix} \begin{bmatrix} \delta x \\ \delta\alpha \\ \delta\nu \end{bmatrix} \end{aligned} \tag{6.41}$$

Substituting for $\delta\nu$ from Equation (6.35) at time t_1 the expansion becomes

$$V = V^0 + (V_x - V_{x\nu}V_{\nu\nu}^{-1}V_\nu)\,\delta x + (V_\alpha - V_{\alpha\nu}V_{\nu\nu}^{-1}V_\nu)\,\delta\alpha$$
$$+ \tfrac{1}{2}[\delta x^{\mathrm{T}}\ \delta\alpha^{\mathrm{T}}]\begin{bmatrix} V_{xx} - V_{x\nu}V_{\nu\nu}^{-1}V_{\nu x} & V_{x\alpha} - V_{x\nu}V_{\nu\nu}^{-1}V_{\nu\alpha} \\ V_{\alpha x} - V_{\alpha\nu}V_{\nu\nu}^{-1}V_{\nu x} & V_{\alpha\alpha} - V_{\alpha\nu}V_{\nu\nu}^{-1}V_{\nu\alpha} \end{bmatrix}\begin{bmatrix}\delta x \\ \delta\alpha\end{bmatrix} \tag{6.42}$$

where all the partial derivatives are evaluated at time t_1.

The problem can now be considered as being without terminal constraints but with a final time equal to t_1. The backward integration of the Equations (6.22) for V'_{xx}, $V'_{\alpha\alpha}$, etc., are initialized with

$$\begin{aligned} V_x'(t_1) &= V_x - V_{x\nu}V_{\nu\nu}^{-1}V_\nu \\ V_\alpha'(t_1) &= V_\alpha - V_{\alpha\nu}V_{\nu\nu}^{-1}V_\nu \\ V'_{\alpha\alpha}(t_1) &= V_{\alpha\alpha} - V_{\alpha\nu}V_{\nu\nu}^{-1}V_{\nu\alpha} \\ V'_{x\alpha}(t_1) &= V_{x\alpha} - V_{x\nu}V_{\nu\nu}^{-1}V_{\nu\alpha} \\ V'_{xx}(t_1) &= V_{xx} - V_{x\nu}V_{\nu\nu}^{-1}V_{\nu x} \end{aligned} \tag{6.43}$$

6.3.3 Free Terminal Time

The final class of problems to be discussed in this chapter is characterized by a free terminal time t_f. The approach employed is similar to that used in deriving the gradient method (see Section 5.2.5) except that second-order expansions are employed.

As before, a change is considered in the final time from t_f to $t_f + dt_f$. The return at the new terminal time $V(x(t_f + dt_f), \alpha, \nu, t_f + dt_f)$ is given by

$$V(x(t_f + dt_f), \alpha, \nu, t_f + dt_f) = \Phi(x(t_f + dt_f), \alpha, \nu, t_f + dt_f) + \int_{t_f}^{t_f+dt_f} L(x, u, \alpha, t)\,dt \tag{6.44}$$

where, for problems with terminal constraints, Φ is given by $\Phi(t_f) = \phi(t_f) + \nu^{\mathrm{T}}\Psi(t_f)$ and, for unconstrained problems, by $\Phi(t_f) = \phi(t_f)$.

The right-hand side is expanded about the nominal time t_f to *second order*

$$V = \Phi(t_f) + [L(x, u, \alpha, t_f) + \Phi_t]\,dt_f + \Phi_x\,\Delta x + \tfrac{1}{2}\{(L + \Phi_{tt})\,dt_f^2 + 2[L_x + \Phi_{tx} + \Phi_{xx}]\,dt_f\,\Delta x + \Delta x^{\mathrm{T}}\Phi_{xx}\,\Delta x\} + \cdots \tag{6.45}$$

where to second order, Δx (see Figure 6.2) is given by

$$\Delta x = f\,dt_f + \tfrac{1}{2}\dot{f}\,dt_f^2 \tag{6.46}$$

and where all the partial derivatives are evaluated at the nominal final time. Substituting for Δx in Equation (6.45) leads to

$$V(t_f + dt_f) = \Phi(t_f) + [L + \Phi_t + \Phi_x f]\, dt_f + \tfrac{1}{2}\{(\dot{L} + \Phi_{tt} + \Phi_x \dot{f}) + [L_x + 2\Phi_{tx}]f + f^{\mathrm{T}}\Phi_{xx} f\}\, dt_f^2 \tag{6.47}$$

Now $V(t_f)$ may be a function of the final state, the control parameters, and the Lagrange multipliers, and variations in these variables must also

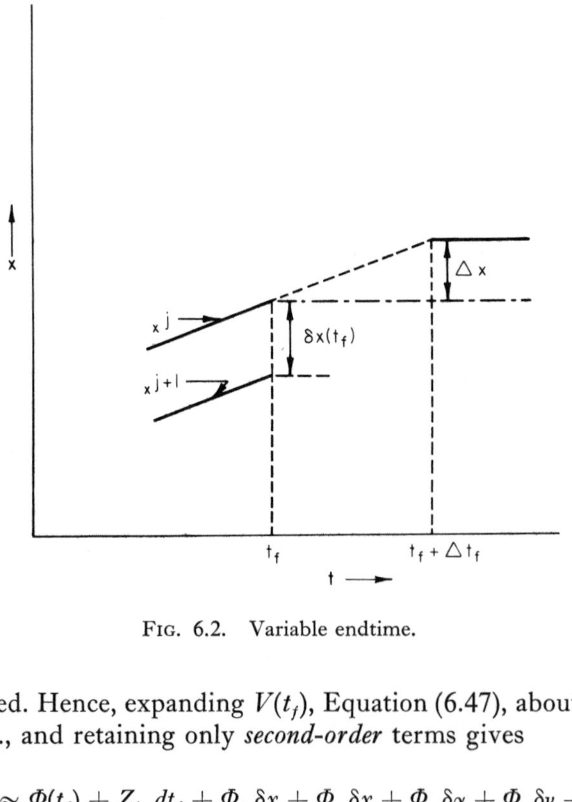

FIG. 6.2. Variable endtime.

be considered. Hence, expanding $V(t_f)$, Equation (6.47), about a nominal state x^j, etc., and retaining only *second-order* terms gives

$$\begin{aligned} V(t_f + dt_f) \approx{}& \Phi(t_f) + Z_{t_f}\, dt_f + \Phi_x\, \delta x + \Phi_\alpha\, \delta x + \Phi_\alpha\, \delta\alpha + \Phi_\nu\, \delta\nu + Z_{\alpha t_f} \delta\alpha\, dt_f \\ &+ Z_{x t_f}\, \delta x\, dt_f + Z_{\nu t_f}\, \delta\nu\, dt_f + \tfrac{1}{2} Z_{t_f t_f}\, dt_f^2 \\ &+ [\delta x^{\mathrm{T}}\ \delta\alpha^{\mathrm{T}}\ \delta\nu^{\mathrm{T}}] \begin{bmatrix} \Phi_{xx} & \text{symmetric} & \\ \Phi_{x\alpha} & \Phi_{\alpha\alpha} & \\ \Phi_{x\nu} & \Phi_{\alpha\nu} & 0 \end{bmatrix} \begin{bmatrix} \delta x \\ \delta\alpha \\ \delta\nu \end{bmatrix} \end{aligned} \tag{6.48}$$

where

$$
\begin{aligned}
Z_{t_f} &= \Phi_t + \Phi_x f + L \\
Z_{xt_f} &= \Phi_{tx} + \Phi_{xx} f + \Phi_x f_x + L_x + (\Phi_x f_u + L_u)\, u_x \\
Z_{\alpha t_f} &= \Phi_{t\alpha} + f^{\mathrm{T}} \Phi_{x\alpha} + \Phi_\alpha f_\alpha + L_\alpha + (\Phi_x f_u + L_u)\, u_\alpha \\
Z_{\nu t_f} &= \Phi_{t\nu} + f^{\mathrm{T}} \Phi_{x\nu} + (\Phi_x f_u + L_u)\, u_\nu
\end{aligned}
\tag{6.49}
$$

and

$$
Z_{t_f t_f} = \Phi_{tt} + 2\Phi_{tx} f + f \Phi_{xx} f + \Phi_x \dot{f} + \dot{L} \tag{6.50}
$$

Two different problems must now be considered: First, the problem *without* terminal constraints and secondly the more general problem *with* terminal constraints. If there are no terminal constraints, dt_f is chosen to maximize the expansion of $V(t_f + dt_f)$, Equation (6.48), i.e.,

$$
dt_f = -Z_{t_f t_f}^{-1} [Z_{t_f} + Z_{t_f x}\, \delta x + Z_{t_f \alpha}\, \delta \alpha] \tag{6.51}
$$

and the new final time, t_f^{j+1} is given by $t_f^{j+1} = t_f^j + dt_f$. Naturally, for a maximum the matrix $Z_{t_f t_f}$ must be negative definite. If the matrix $Z_{t_f t_f}$ is positive or has less than full rank, a gradient correction should be used.

The variation in the final time may now be eliminated from the expansion of $V(t_f + dt_f)$, Equation (6.48), which becomes

$$
\begin{aligned}
V(t_f + dt_f) = \Phi(t_f) &- Z_{t_f} Z_{t_f t_f} Z_{t_f} + (\Phi_x - Z_{t_f} Z_{t_f t_f}^{-1} Z_{t_f x})\, \delta x \\
&+ (\Phi_\alpha - Z_{t_f} Z_{t_f t_f}^{-1} Z_{t_f \alpha})\, \delta \alpha \\
&+ [\delta x^{\mathrm{T}}\ \delta \alpha^{\mathrm{T}}]
\begin{bmatrix}
\Phi_{xx} - Z_{t_f x}^{\mathrm{T}} Z_{t_f t_f}^{-1} Z_{t_f x} & \text{symmetric} \\
\Phi_{\alpha x} - Z_{t_f \alpha}^{\mathrm{T}} Z_{t_f t_f}^{-1} Z_{t_f x} & \Phi_{\alpha\alpha} - Z_{t_f \alpha}^{\mathrm{T}} Z_{t_f t_f}^{-1} Z_{t_f \alpha}
\end{bmatrix}
\begin{bmatrix} \delta x \\ \delta \alpha \end{bmatrix}
\end{aligned}
\tag{6.52}
$$

Thus the terminal conditions for the partial derivatives of the return function become

$$
\begin{aligned}
V_x(t_f) &= \Phi_x - Z_{t_f} Z_{t_f t_f}^{-1} Z_{t_f x} \\
V_\alpha(t_f) &= \Phi_\alpha - Z_{t_f} Z_{t_f t_f}^{-1} Z_{t_f \alpha} \\
V_{\alpha x}(t_f) &= \Phi_{\alpha x} - Z_{\alpha t_f} Z_{t_f t_f}^{-1} Z_{t_f x} \\
V_{xx}(t_f) &= \Phi_{xx} - Z_{x t_f} Z_{t_f t_f}^{-1} Z_{t_f x} \\
V_{\alpha\alpha}(t_f) &= \Phi_{\alpha\alpha} - Z_{\alpha t_f} Z_{t_f t_f}^{-1} Z_{t_f \alpha}
\end{aligned}
\tag{6.53}
$$

In the second type of problem, when terminal constraints are present, the variation in the final time should be eliminated by using one of the side constraints, say ψ_k, as a *stopping condition*. On the forward integration t_f^{j+1} is chosen such that ψ_k is satisfied or nearly satisfied. The multiplier associated with ψ_k, ν_k is chosen by noting that for an optimum solution

$$Z_{t_f} = 0 = \Phi_t + \Phi_x f + L$$

where $\Phi = \phi(x(t_f)) + \nu^T\psi$ and

$$\nu = [\nu_1, \nu_2, \ldots, \nu_k, \ldots]^T$$

Thus

$$\nu_k = (1/\dot{\psi}_k)(\dot{\phi} + \nu'\dot{\psi}' + L)$$

where

$$\nu' = [\nu_1, \nu_2, \ldots, \nu_{k-1}, \nu_{k+1}, \ldots]^T$$

and

$$\psi' = [\psi_1, \psi_2, \ldots, \psi_{k-1}, \psi_{k+1}, \ldots]^T$$

The variation in the final time dt_f and the kth multiplier may now be eliminated by chosing them to maximize the expansion equation (6.48). The variations dt_f, $\delta\nu_k$ are given by

$$\begin{bmatrix} dt_f \\ \delta\nu_k \end{bmatrix} = -\begin{bmatrix} Z_{t_f t_f} & Z_{t_f \nu_k} \\ Z_{t_f \nu_k} & 0 \end{bmatrix}^{-1} \left[\begin{pmatrix} 0 \\ Z_{\nu_k} \end{pmatrix} + \begin{pmatrix} Z_{t_f x} \\ Z_{\nu_k x} \end{pmatrix} \delta x + \begin{pmatrix} Z_{t_f \alpha} \\ Z_{\nu_k \alpha} \end{pmatrix} \delta\alpha + \begin{pmatrix} Z_{t_f \nu'} \\ 0 \end{pmatrix} \delta\nu' \right] \quad (6.54)$$

(**Note:** If the stopping condition was satisfied exactly, Z_{ν_k} is, of course, identically zero.) These expressions may now be used to eliminate dt_f and $\delta\nu_k$. The terminal conditions for the partial derivatives are again modified, and the evaluation of these conditions is left to the reader as an exercise.

6.4 EXAMPLES

EXAMPLE 6.1. First, the brachistochrone problem, Example 5.2, will be considered. This will enable a comparison to be made with the other methods. The problem was described in detail in Chapter 3, and only the relevant equations are included here.

The system equation is

$$\dot{x} = u, \qquad x(0) = 0 \quad (6.55)$$

and the performance index (to be minimized) is J where

$$J = \int_0^1 (1 + u^2)^{1/2}/(1 - x)^{1/2}\, dt \tag{6.56}$$

The terminal constraint is $\psi = x(1) + 0.5 = 0$. The partial derivatives needed to implement the successive sweep method are

$$L_x = 1/2\sqrt{1 + u^2}/(1 - x)^{3/2}, \qquad L_u = u\sqrt{1 + u^2}/(1 - x)^{1/2}$$

$$f_x = 0, \qquad f_u = 1, \qquad f_{xx} = f_{uu} = f_{xu} = 0$$

$$L_{xx} = 3/4(1 + u^2)^{1/2}/(1 - x)^{5/2}$$

$$L_{xu} = 1/2u(1 + u^2)^{-1/2}(1 - x)^{-3/2}$$

$$L_{uu} = (1 - x)^{-1/2}(1 + u^2)^{-3/2}$$

The boundary conditions for the partial derivatives required in the backwards computation are

$$V_x = \nu, \qquad V_\nu = x + 0.5, \qquad V_{xx} = V_{x\nu} = V_{\nu\nu} = 0$$

This completes the relations needed for the algorithm. The procedure used may now be summarized as follows.

1. The state equation (6.55) was integrated forward using the nominal control $u(s) = 0$, $0 \leqslant s \leqslant 0.5$, $u(s) = -1$, $0.5 \leqslant s \leqslant 1$. The state was stored at each integration step and the cost J^j was also computed.

2. The equations for the partial derivatives of V, i.e., V_x, V_ν, V_{xx}, $V_{x\nu}$, and $V_{\nu\nu}$, were integrated backward in time using the stored values of the state. At the same time the gradient $V_x + u/(1 + u^2)^{1/2}(1 - x)^{1/2}$, Z_{uu} and the feedback gains $Z_{uu}^{-1}Z_{ux}$, $Z_{uu}^{-1}Z_{u\nu}$ were stored at each integration step.

3. At the initial time the change in ν, $\delta\nu$ was computed [cf Equation (6.44)], and the new multiplier formed from $\nu^{j+1} = \nu^j + \delta\nu$.

4. The state equation was integrated forward using the new control u^{j+1} where $u^{j+1} = u^j + \delta u$ with δu given by

$$\delta u = -Z_{uu}^{-1}\{Z_u + Z_{ux}\,\delta x + Z_{u\nu}\,\delta\nu\}$$

and $\delta x = x^{j+1} - x^j$. A step size parameter was not needed.

5. Steps 2–4 were repeated until no further reduction in the cost was obtained.

The convergence is shown in Table XXI and should be compared with Table XVII for the gradient method. As can be seen, the optimal

TABLE XXI. CONVERGENCE FOR EXAMPLE 6.1

Iteration Number	Cost	$u(0)$	$x(1)$	ν
Nominal	1.1356745	−0.0000000	−0.5000000	0.0000000
1	1.0009902	−0.6730697	−0.5032761	0.3495743
2	0.9985866	−0.7789593	−0.5003774	0.2222272
3	0.9985008	−0.7834336	−0.5000088	0.2193740
4	0.9984989	−0.7834292	−0.5000000	0.2193624

cost is obtained with about half the number of iterations. The state and control histories are very similar to those obtained for the discrete case and so are not repeated.

The equations were integrated with a standard fourth-order Runge–Kutta scheme with a step length of 0.01. On the backward integration the stored values of the state were linearly interpolated where necessary. The computer program took approximately 2 sec per iteration including printout.

EXAMPLE 6.2. Next the previous example will be considered with a variable end time and a slightly different terminal constraint. The state equation (6.55) remains unaltered, but the performance criterion becomes

$$J = \int_0^{t_f} \{(1+u^2)^{1/2}/(1-x)^{1/2}\}\, dt \tag{6.57}$$

where the final time t_f is free. The terminal constraint becomes

$$\psi_1 = x(t_f) + 0.5 - t_f = 0 \tag{6.58}$$

The constraint will be used as a stopping condition to eliminate variations in the final time. Since ν is eliminated at the final time, the partials V_ν, $V_{x\nu}$, and $V_{\nu\nu}$ need not be computed. The equations for V_x and V_{xx} remain unaltered except those for the terminal conditions, which are given by (cf Section 6.2.3)

$$\begin{aligned} V_x(t_f) &= Z_x \\ V_{xx}(t_f) &= 2Z_x/(u-1) + Z_x u/(u-1)^2 \end{aligned} \tag{6.59}$$

where $Z_x = 0.5(1+u^2)^{1/2}/(1-x)^{3/2}$.

The successive sweep procedure may now be summarized briefly as follows.

1. The state equations are integrated forward until the stopping condition, Equation (6.58), is satisfied using the nominal control $u^j(s) = -1, 0 \leqslant s \leqslant t_f$.

2. At the final time the boundary values for V_x and V_{xx} are computed (cf Section 6.3.3).

3. The equations for V_x and V_{xx} are integrated backward and the function $Z_{uu}^{-1}Z_u$ and the gain $Z_{uu}^{-1}Z_{ux}$ are computed and stored at each integration step.

4. The state equation is integrated forward using the new nominal u^{j+1} where $u^{j+1} = u^j - Z_{uu}^{-1}(Z_u + Z_{ux}\,\delta x)$. Again no step length control was required.

5. Steps 2, 3, and 4, were repeated until no further reduction in cost was obtained.

A summary of the convergence is given in Table XXII. As can be seen, only 6 iterations are required, which is a substantial reduction when compared with the gradient method (16 iterations, Table XVIII). The optimal trajectories are very similar to those obtained with the gradient technique (Figure 5.3) and so will not be repeated here. The computation details are similar to those in the previous example.

TABLE XXII. CONVERGENCE FOR EXAMPLE 6.2

Iteration Number	Cost	$u(1)$	$x(t_f)$	t_f
Nominal	0.3338505	−1.0000000	−0.2500000	0.2500000
1	0.3332400	−1.2055273	−0.2627421	0.2372579
2	0.3332369	−1.2336984	−0.2633844	0.2366156
3	0.3332370	−1.2346277	−0.2633977	0.2366023
4	0.3332370	−1.2346786	−0.2633997	0.2366003
5	0.3332370	−1.2346790	−0.2633996	0.2366004
6	0.3332370	−1.2346792	−0.2633996	0.2366004

EXAMPLE 6.3. *Aircraft Landing Problem.* In this and the following example the successive sweep method is applied to Examples 5.4 and 5.5. The setting up of the aircraft landing problem as an optimal control problem was fully discussed in Example 5.4 and will not be repeated here.

It may be recalled that the problem was to minimize the performance index J where

$$J = A(x_4 - 8)^2 + \int_0^{60} \{A_1(x_1 - u_D)^2 + A_2(x_5 - h_D)^2 + g_1 u_1^2 + g_2(u_2 - 0.75)^2\}\, dt \tag{6.60}$$

and where the weighting parameters A, A_1, A_2, g_1, and g_2 were chosen to be

$$A = A_1 = A_2 = 1, \qquad g_1 = 0.33, \qquad g_2 = 20 \tag{6.61}$$

The state equations were

$$\begin{aligned}
\dot{x}_1 &= -0.041x_1 + 0.381x_2 - 0.562x_3 - 2.552x_7 + 0.221R \\
\dot{x}_2 &= -0.066x_1 - 0.577x_2 + x_3 - 0.050x_6 - 0.992x_7 - .395R \\
\dot{x}_3 &= 0.011x_1 - 1.108x_2 - 0.822x_3 - 1.264x_6 - 0.154x_7 - 3.544R \\
\dot{x}_5 &= -12.147 + 4.049(x_4 - x_2) \\
\dot{x}_6 &= u_1 \\
\dot{x}_7 &= (u_2 - x_7)/3
\end{aligned} \tag{6.62}$$

where the seven elements of the state vector $\mathbf{x}$ are as follows.

x_1 = increment of forward speed (u_A, ft/sec)

x_2 = increment of altitude angle (α, deg)

x_3 = increment of pitch rate ($\dot{\theta}$, deg/sec)

x_4 = increment of pitch angle (θ, deg)

x_5 = height (h, ft)

x_6 = increment of elevator angle (η, deg)

x_7 = increment of throttle (Th)

u_1 = increment of elevator rate (deg/sec)

u_2 = input to throttle actuator

and where R, the ground effect term, is $R = 400/(3x_5 + 100) - 1$.

The matrices f_x, f_u, etc., required to implement the successive sweep procedure are as follows.

$$f_x = \begin{bmatrix} -0.041 & 0.381 & 0.000 & -0.562 & 0.221*R_x & 0.000 & -2.552 \\ -0.066 & -0.577 & 1.000 & 0.000 & -0.395*R_x & -0.050 & 0.992 \\ 0.011 & -1.108 & -0.822 & 0.000 & -3.607*R_x & -1.264 & -0.157 \\ 0.000 & 0.000 & 1.000 & 0.000 & 0.000 & 0.000 & 0.000 \\ 0.000 & -4.049 & 0.000 & 4.049 & 0.000 & 0.000 & 0.000 \\ 0.000 & 0.000 & 0.000 & 0.000 & 0.000 & 0.000 & 0.000 \\ 0.000 & 0.000 & 0.000 & 0.000 & 0.000 & 0.000 & -0.333 \end{bmatrix}$$

$$f_u = \begin{bmatrix} 0 & 0 \\ 0 & 0 \\ 0 & 0 \\ 0 & 0 \\ 0 & 0 \\ 1 & 0 \\ 0 & \frac{1}{3} \end{bmatrix}, \qquad f_{xx} = \begin{bmatrix} 0 & 0 & 0 & 0 & 0.221*R_{xx} & 0 & 0 \\ & & & 0 & -0.395*R_{xx} & 0 & 0 \\ & & & 0 & -3.607*R_{xx} & 0 & 0 \\ & & & 0 & 0 & 0 & 0 \\ & & & 0 & 0 & 0 & 0 \\ & & & 0 & 0 & 0 & 0 \\ & & & 0 & 0 & 0 & 0 \end{bmatrix}$$

$$L_x = \begin{bmatrix} u_Z{}^i - u_D \\ 0 \\ 1 \\ h^i - h_D \\ 0 \\ 0 \\ 0 \end{bmatrix}, \qquad L_u = \begin{bmatrix} 2g_1u_1 \\ 2g_2(u_2 - 0.75) \end{bmatrix}$$

$$L_{ux} = 0, \qquad L_{uu} = 0, \qquad L_{xx} = \begin{bmatrix} 1.0 & 0.0 & 0.0 & 0.0 & 0.0 & 0.0 & 0.0 \\ 0.0 & 0.0 & 0.0 & 0.0 & 0.0 & 0.0 & 0.0 \\ 0.0 & 0.0 & 1.0 & 0.0 & 0.0 & 0.0 & 0.0 \\ 0.0 & 0.0 & 0.0 & 0.0 & 0.0 & 0.0 & 0.0 \\ 0.0 & 0.0 & 0.0 & 0.0 & 1.0 & 0.0 & 0.0 \\ 0.0 & 0.0 & 0.0 & 0.0 & 0.0 & 0.0 & 0.0 \\ 0.0 & 0.0 & 0.0 & 0.0 & 0.0 & 0.0 & 0.0 \end{bmatrix}$$

where R_x equals $-1200/(3x_5 + 100)^2$ and R_{xx} equals $-7200/(3x_5 + 100)^3$. The initial conditions for the backward integration are $V_x(t_f) = [0, 0, 0, x_4 - 8, 0, 0, 0]$ and $V_{xx}(t_f) = 0$ except $V_{x_4x_4}(t_f) = 1$. The successive sweep method was then used to compute the optimal control.

However, before discussing the results it is of interest to discuss the choice of a priming trajectory. Actually, the priming trajectory has little or no effect on the convergence but, in this case, it was possible to find a very good priming trajectory. The reader may have observed that the state equations (6.62) contain only one nonlinear term R and that this nonlinearity is a function of a single variable x_5, the height. As any near optimal height trajectory should follow the desired trajectory h_D closely, it is reasonable to suppose that a very good approximation to the system equations might be obtained by linearizing the terms in R about h_D.

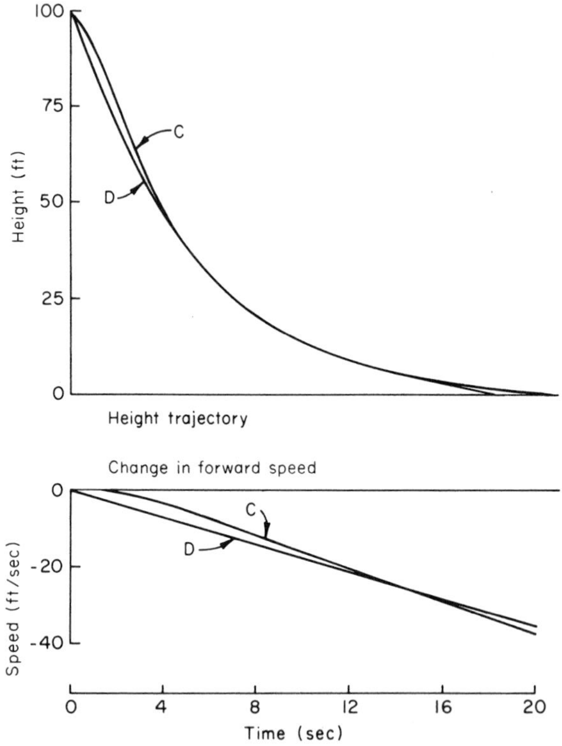

FIG. 6.3a. Optimal landing trajectories. D is the ideal trajectory, C the optimal trajectory.

Expanding R to first order gives $R(x_5) = R(h_D) + (\partial R/\partial x_5)\,\Delta x$ where $\Delta x = x_5 - h_D$. Thus the term R may be replaced, at least to first order, by $R(x_5) = \{R(h_D) - R_{x_5}(h_D)\,h_D\} - R_{x_5}(h_D)\,x_5$. The approximated system equations are *linear* and, as the performance index is quadratic, the optimal control u^* for this *new system* may be computed using the algorithm described in Chapter 4. The control u^* was then used as a priming solution for the nonlinear problem.

This technique can only be used if the system nonlinearity can be suitably linearized. However, it does reduce the number of iterations required by giving a good priming solution. For problems with a large number of state variables this can represent a substantial saving in computer time.

The successive sweep algorithm was then used to compute the optimal control law and the optimal state trajectory (see Figure 6.3).

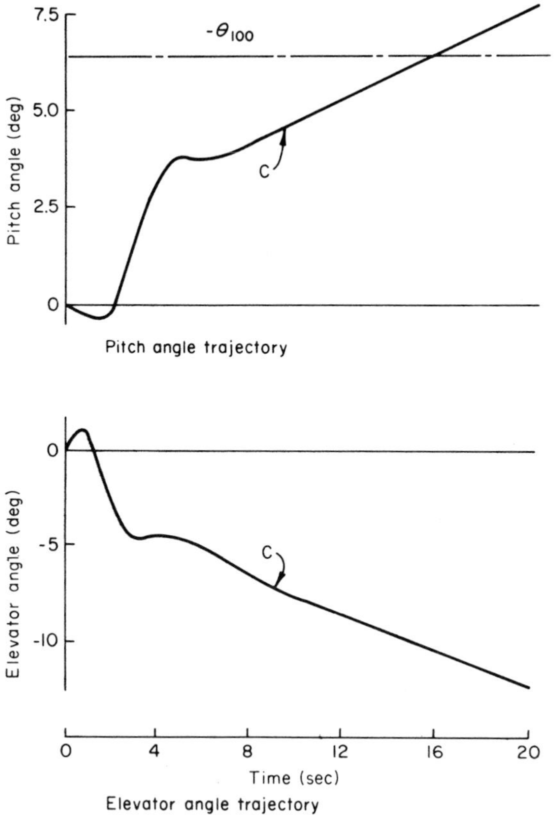

FIG. 6.3b. Elevator angle trajectory.

As can be seen, the height trajectory does in fact follow the desired trajectory very closely; the difference is less than six inches for the greater part of the landing. Convergence was obtained in only four iterations, as is shown in Table XXIII. The convergence was much faster than with the gradient method. Furthermore, the cost, 221, is less than half that computed with the gradient procedure. Clearly the gradient solution was suboptimal.

TABLE XXIII

CONVERGENCE OF SUCCESSIVE SWEEP ALGORITHM

Iteration Number	Cost	Convergence Factor, ϵ
Priming	3514	
1	489	1.0
2	265	1.0
3	222	1.0
4	221	1.0

The computation of the optimal control took approximately 60 sec per iteration on the IBM 7090 computer and used about 17,000 words of computer storage.

EXAMPLE 6.4. *Control of a Boiler.* As a final example, the successive sweep algorithm is applied to the problem discussed in Example 5.5. The formulation of the boiler problem as an optimal control problem was fully discussed in Example 5.5 and so it will not be repeated.

The problem was to minimize the performance index J

$$J = \int_0^{150} \{(x_2 - 97920)^2 + 4 \times 10^2 x_5 + 10^2 u^2\}\, dt \tag{6.63}$$

where x_2 equals the drum pressure, x_5 equals the fuel flow rate, and u equals the input to the actuator controlling the fuel flow and where the state equations were given in Equation (5.63).

The matrices f_x, f_{xx}, etc., required to implement the successive sweep procedure may now be evaluated. These partial derivatives are somewhat lengthy and their computation is left to the reader as an exercise.

The application of the algorithm is then quite straightforward although the choice of the step-length parameter ϵ was fairly important. The convergence for various step parameters is shown in Table XXIV. A step size of unity is not generally successful, but, as Table XXIV shows, the use of smaller basic convergence factors is even less successful. Therefore, despite the frequent reduction in ϵ, the policy of commencing each iteration with $\epsilon = 1$ and halving this if necessary was adopted.

The effect of the priming trajectory on the convergence is not very marked. Table XXV shows the convergence of cost from three priming trajectories.

TABLE XXIV

EFFECT OF CONVERGENCE FACTOR ON RATE OF CONVERGENCE[a]

Iteration	ϵ	Cost $\times 10^{-9}$	ϵ	Cost $\times 10^{-9}$	ϵ	Cost $\times 10^{-9}$
0		1.726585		1.726585		1.726585
1	1.0	1.441430	0.8	1.440981	0.6	1.468406
2	0.5	1.424152	0.8	1.424447	0.6	1.429297
3	0.5	1.421580	0.8	1.423203	0.6	1.426844
4	0.5	1.421346	0.8	1.423106	0.6	1.422160
5	1.0	1.421311	0.4	1.421490	0.6	1.421469
6	0.5	1.421298	0.8	1.421467	0.6	1.421325
7	0.5	1.421294	0.4	1.421308	0.6	1.421303
8	0.125	1.421294	0.4	1.421295	0.6	1.421298

[a] Step change in steam load, 90 to 100 percent full load.

TABLE XXV

EFFECT OF PRIMING TRAJECTORY ON RATE OF CONVERGENCE[a]

	Primed $u(t) = 90\%$		Primed $u(t) = 100\%$		Primed Approx. Optimal	
Iteration	ϵ	Cost $\times 10^{-9}$	ϵ	Cost $\times 10^{-9}$	ϵ	Cost $\times 10^{-9}$
0		1.151793		1.726585		1.433536
1	1.0	1.486317	1.0	1.441430	0.25	1.431088
2	1.0	1.442335	0.5	1.424152	0.5	1.427843
3	0.25	1.434390	0.5	1.421580	1.0	1.427709
4	1.0	1.433496	0.5	1.421346	0.25	1.425913
5	0.25	1.429428	1.0	1.421311	1.0	1.423028
6	1.0	1.427189	0.5	1.421298	0.25	1.422433
7	0.25	1.425157	0.5	1.421294	0.125	1.422317
8	0.5	1.422415	0.125	1.421294		

[a] Step change in steam load, 90 to 100 percent full load.

1. The control is left at its previous steady-state value.

2. The control is increased to the new steady-state value (that is, immediate steady-state optimization).

3. The control is a step-wise approximation to the optimum trajectory.

It is clear from this table that priming procedure 2 gives the best convergence, although the algorithm is able to improve all of the priming trajectories quite satisfactorily.

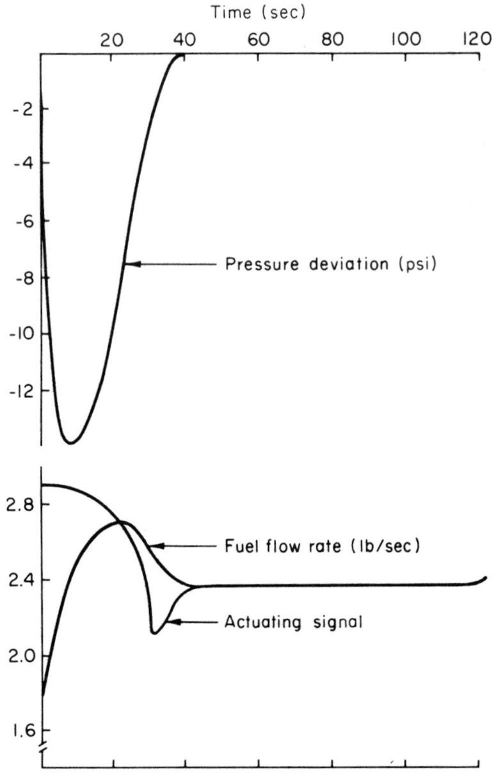

FIG. 6.4. Optimal control. (Step change in steam load, 90 to 100 percent full load.)

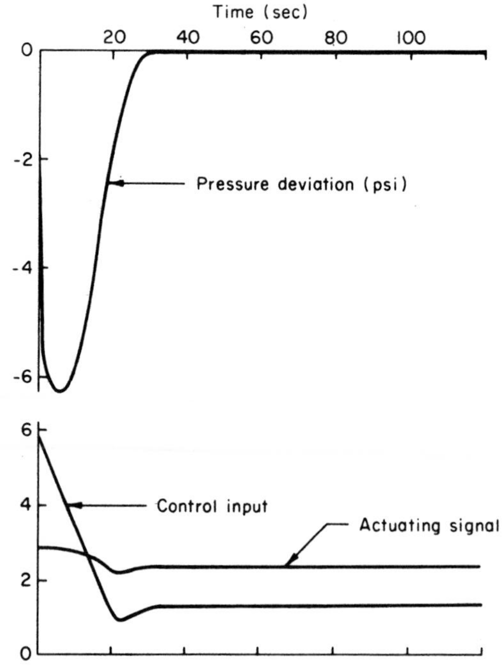

FIG. 6.5. Optimal control. (Step change in steam load, 80 to 100 percent full load.)

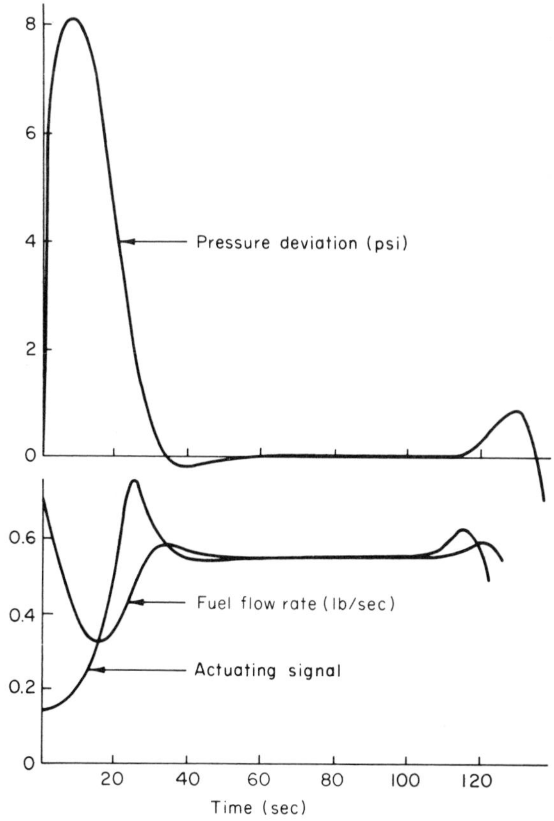

FIG. 6.6. Optimal control. (Step change in steam load, 40 to 30 percent full load.)

Tables XXIV and XXV also show that for this particular disturbance, a step change in steam load from 90 to 100 percent full load, only eight iterations are required to determine the optimum trajectories (Figure 6.4). Some optimal trajectories for some other disturbances are shown in Figures 6.5 and 6.6. The maximum pressure error is small, even on the severe disturbance from 80 to 100 percent, and the overshoot is negligible. The computation time per iteration was about 56 sec on the IBM 7090 computer, as compared to 14 sec for the gradient method. However, the optimal solution was found in less than ten iterations whereas after 100 iterations of the gradient method the control was still far from optimal. This computation time is still too large for on-line implementation with present-day computers, although improved programming would help to reduce the time per iteration. Even so the method is valuable for off-line optimization and to provide a basis for the comparison of suboptimal systems.

6.5 NEIGHBORING OPTIMAL CONTROL

A very useful by-product of the successive sweep algorithm relates to the technique of neighboring optimal control (cf Section 3.8). The technique of neighboring control is based on an approximation to the optimal control law in a region of state space which is in the neighborhood of the optimal trajectory. Such approximations are useful for generating new optimal solutions and for studying optimal guidance schemes.

To illustrate the technique, the problem without control parameters, without side constraints, and with a fixed terminal time will be considered. In other words, the problem is to maximize the performance index J where,

$$J = \phi(x(t_f), t_f) + \int_{t_0}^{t_f} L(x, u, t)\, dt \tag{6.64}$$

and where

$$\dot{x} = f(x, u, t), \qquad x(t_0) = x_0 \text{ specified}, \qquad t_f \text{ specified} \tag{6.65}$$

It is assumed that the open-loop optimal solution $u^{\text{opt}}(t)$ (and hence $x^{\text{opt}}(t)$) is known. Then the neighboring control method seeks the following approximation to the control law

$$u^{\text{opt}}(x, t) \approx u^{\text{opt}}(t) + \frac{\partial u^{\text{opt}}}{\partial x}(x^{\text{opt}}(t), t)[x(t) - x^{\text{opt}}(t)] \tag{6.66}$$

Now, it may be recalled that as the successive sweep method converges, $\delta u^0 \to 0$ and the new nominal control u^{j+1} becomes

$$u^{j+1} = u^j - Z_{uu}^{-1} Z_{ux}\, \delta x \tag{6.67}$$

where Z_{uu} and Z_{ux} are determined from Equation (6.5). Clearly, the two control laws, Equations (6.56) and (6.57), are identical if

$$\partial u^{\text{opt}}/\partial x = -Z_{uu}^{-1} Z_{ux} \tag{6.68}$$

That the equations are identical may be shown easily. Along the optimal trajectory the stationarity condition, Equation (5.66), must hold, i.e.,

$$V_x^{\text{opt}} f_u + L_u = 0 \tag{6.69}$$

The partial differentiation of this equation with respect to x gives

$$V_{xx}^{\text{opt}} f_u + V_x^{\text{opt}} f_{ux} + L_{ux} + (V_x^{\text{opt}} f_{uu} + L_{uu}) \frac{\partial u^{\text{opt}}}{\partial x} = 0 \tag{6.70}$$

Thus, providing the coefficient of $\partial u^{\text{opt}}/\partial x$, $V_x^{\text{opt}} f_{uu} + L_{uu}$, is nonsingular,

$$\partial u^{\text{opt}}/\partial x = -[V_x^{\text{opt}} f_{uu} + L_{uu}]^{-1}[V_{xx}^{\text{opt}} f_u + V_x^{\text{opt}} f_{ux} + L_{ux}] \tag{6.71}$$

This is exactly the same feedback law as that produced by the successive sweep algorithm provided V_x^{opt}, V_{xx}^{opt} can be identified with V_x^j, V_{xx}^j [cf Equation (6.11)].

Once convergence has been obtained, the equations for V_x^j, V_{xx}^j become

$$\begin{aligned} V_x^j(t_f) &= \phi_x(t_f) \\ \dot{V}_x^j &= -V_x^j f_x - L_x \end{aligned} \tag{6.72}$$

$$\begin{aligned} V_{xx}^j(t_f) &= \phi_{xx}(t_f) \\ \dot{V}_{xx}^j &= -Z_{xx} - Z_{xu} Z_{uu}^{-1} Z_{ux} \end{aligned} \tag{6.73}$$

Now the characteristic equations for V_x^{opt}, V_{xx}^{opt} are obtained by taking partial derivatives of the dynamic programming equation for V^{opt}

$$\begin{aligned} V^{\text{opt}}(x(t_f), t_f) &= \phi(x(t_f), t_f) \\ V_t^{\text{opt}} + V_x^{\text{opt}} f(x, u^{\text{opt}}, t) + L(x, u^{\text{opt}}, t) &= 0 \end{aligned} \tag{6.74}$$

Taking the partial derivative of Equations (6.74) with respect to x yields $V_x^{\text{opt}}(t_f) = \phi_x(t_f)$ and

$$V_{tx}^{\text{opt}} + V_{xx}^{\text{opt}} f + V_x^{\text{opt}} f_x + L_x + (V_x^{\text{opt}} f_u + L_u)\frac{\partial u^{\text{opt}}}{\partial x} = 0 \tag{6.75}$$

However, since the stationarity condition, Equation (6.69), holds, this equation may be simplified to

$$\dot{V}_x^{\text{opt}} = -V_x^{\text{opt}} f_x - L_x \tag{6.76}$$

Now, since Equation (6.76) is identical to Equation (6.72) and $V_x^{\text{opt}}(t_f)$ is identical to $V_x^j(t_f)$,

$$V_x^j(t) = V_x^{\text{opt}}(t) \qquad t \text{ in } [t_0, t_f] \tag{6.77}$$

The equivalence of V_{xx}^{opt} and V_{xx}^j may be shown in a similar manner by taking partial derivatives of Equations (6.75),

$$V_{xx}^{\text{opt}}(t_f) = \phi_{xx}(t_f), \qquad \dot{V}_{xx}^{\text{opt}} = -Z_{xx} - Z_{xu} Z_{uu}^{-1} Z_{ux} \tag{6.78}$$

This establishes the desired result.

One application of this analysis may be illustrated easily in terms of one of the previous examples, 6.3, the aircraft landing problem. For, during the landing the aircraft is subjected to many small disturbances. One approach to the problem of handling such disturbances is to use a local feedback law as just derived. The mechanization of such a system is shown in Figure 6.7, and some results are shown in Table XXVI. The

TABLE XXVI. PARAMETERS AT TOUCHDOWN

Approach Speed	Landing Time (sec)	Forward Speed ($>1.1V_s$)	Rate of Descent $\dot{h}$ (ft/sec)	Overall Pitch (deg)
Nominal	20.5	−2.2	−0.8	1.2
+20 ft/sec	20.5	−3.1	−0.7	1.2

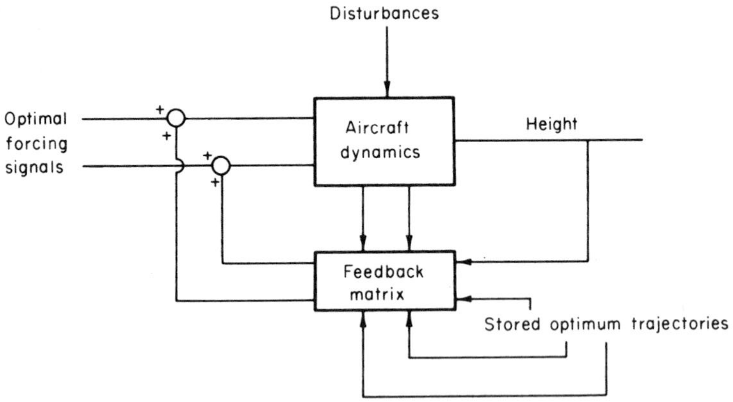

FIG. 6.7. Mechanization of the neighboring optimal controller.

first row of figures contains the nominal touchdown parameters using the optimal control. The second row contains these same parameters when the initial speed is perturbed by 20 ft/sec and using the neighboring feedback law just discussed. As can be seen in this simple case, this feedback law was quite adequate. Dyer (1966) examined a variety of perturbations for this problem and found that the neighboring feedback law gave good performance even when gross disturbances were added.

6.6 THE SECOND VARIATION AND THE CONVEXITY CONDITION

From the derivation of the successive sweep method it follows that the convexity condition for a local optimum may be expressed as the condition

$$V_{xx}^{\text{opt}} f_{uu} + L_{uu} \leqslant 0, \qquad 0 \leqslant t \leqslant t_f \tag{6.79}$$

A further derivation will be given which is based on the second variation of the return function $\delta^2 V(x(t_0), t_0)$.

As in Section 5.4, let $u^0(t)$ denote some nominal control and $V^0(x, t)$ the related return function, i.e.,

$$V^0(x_0, t_0) = \phi(x(t_f), t_f) + \int_{t_0}^{t_f} L(x, u^0 t)\, dt \tag{6.80}$$

with $x(t_0) = x_0$ specified and $\dot{x}$ equals $f(x, u^0, t)$. Let $u'(t)$ denote a control "close" to $u^0(t)$, i.e.,

$$u'(t) = u^0(t) + \epsilon\eta(t) \tag{6.81}$$

where $|\epsilon| \ll 1$ i.e., such that $\| \delta u \| = \| u' - u^0 \| \leqslant \delta$.

Now in Section 5.4 it was shown that the difference between the return related to the nominal control u^0 and the return related to the control u' may be expressed as

$$\delta V(t_0) = \int_{t_0}^{t_f} \{V_t^0(x', t) + V_x^0(x', t) f(x', u', t) + L(x', u', t)\}\, dt \tag{6.82}$$

where x' is the trajectory defined by the control u'. Following the analysis of Section 5.4, this expression is expanded about the control u^0 and state x^0 giving to second order

$$\begin{aligned} \delta V = \int_{t_0}^{t_f} &\{[V_{tx}^{c} + V_{xx}^0 f + V_x^0 f_x + L_x]\, \delta x + [V_x^0 f_u + L_u]\, \delta u \\ &+ 1/2 \delta x^{\mathrm{T}} [V_{txx}^0 + V_{xxx}^0 f + V_{xx}^0 f_x + f_x^{\mathrm{T}} V_{xx}^0 + V_x^0 f_{xx} + L_{xx}]\, \delta x \\ &+ \delta x^{\mathrm{T}} [V_{xx}^0 f_u + V_x^0 f_{xu} + L_{xu}]\, \delta u \\ &+ \delta u^{\mathrm{T}} [V_x^0 f_{uu} + L_{uu}]\, \delta u\}\, dt \end{aligned} \tag{6.83}$$

It was shown in Section 5.4 that the first-order terms referred to as the first variation $\delta' V$, become $\delta' V = \epsilon \int_{t_0}^{t_f} \{V_x^0 f_u + L_u\}\, \eta\, dt$.

The second variation is derived from the remaining second-order terms in the expansion. First, the variation δx must be expressed in terms of the change in the control $u' - u^0$. Expanding δx in terms of ϵ gives, to second order,

$$\delta x = \epsilon \beta(t) + O(\epsilon^2) \tag{6.84}$$

where $\beta(t)$ may be found by expanding the state equation about $\dot{x}$ and $\dot{u}$, viz,

$$\dot{x}^1 = \dot{x}^0 + \epsilon\dot{\beta} + O(\epsilon^2) = f(x^0, u^0, t) + \epsilon[f_x\beta + f_u\eta] + O(\epsilon^2)$$

Hence, $\dot{\beta}(t) = f_x\beta + f_u\eta$. Thus, substituting Equation (6.84) for δx leads to the following expression for the second variation $\delta^2 V$

$$\delta^2 V = \tfrac{1}{2}\epsilon^2 \int_{t_0}^{t_f} \{\beta^{\mathrm{T}}[V_{txx}^0 + V_{xxx}^0 f + V_{xx}^0 f_x + f_x{}^{\mathrm{T}} V_{xx}^0 + V_x{}^0 f_{xx} + L_{xx}]\,\beta$$
$$+ 2\beta^{\mathrm{T}}[V_{xx}^0 f_u + V_x{}^0 f_{xu} + L_{xu}]\,\eta + \eta^{\mathrm{T}}[V_x^0 f_{uu} + L_{uu}]\,\eta\}\,dt \tag{6.85}$$

Equation (6.85) may be simplified by noting that

$$\int_{t_0}^{t_f} \{\delta\beta^{\mathrm{T}}(V_{txx}^0 + V_{txxx}^0 f)\,\delta\beta + \delta\beta^{\mathrm{T}} V_{xx}^0[f_x\,\delta\beta + f_u\,\delta\eta]$$
$$+ (f_x\,\delta\beta + f_u\,\delta\eta)^{\mathrm{T}} V_{xx}^0\,\delta\beta\}\,dt$$
$$= \int_{t_0}^{t_f} \{\delta\beta^{\mathrm{T}}\dot{V}_{xx}^0\,\delta\beta + \delta\beta V_{xx}^0\,\delta\dot{\beta} + \delta\dot{\beta} V_{xx}^0\,\delta\beta\}\,dt$$

which may be written

$$\int_{t_0}^{t_f} \left\{\frac{d(\delta\beta^{\mathrm{T}} V_{xx}^0\,\delta\beta)}{dt}\right\} dt = \delta\beta\phi_{xx}\,\delta\beta\,|_{t=t_f} \tag{6.86}$$

(This assumes that the initial state is fixed, i.e., $\delta\beta(t_0) = 0$.)

Thus the second variation may be written

$$\delta^2 V = \epsilon^2/2[\delta\beta(t_f)^{\mathrm{T}}\phi_{xx}(x(t_f), t_f)\,\delta\beta(t_f) + \int_{t_0}^{t_f} \{\delta\beta^{\mathrm{T}}[V_x{}^0 f_{xx} + L_{xx}]\,\delta\beta$$
$$+ 2\delta\beta^{\mathrm{T}}[V_x{}^0 f_{xu} + L_{xu}]\,\delta\eta + \delta\eta^{\mathrm{T}}[V_x{}^0 f_{uu} + L_{uu}]\,\delta\eta\}\,dt \tag{6.87}$$

A similar expression for the second variation may be obtained via the classical approach. In Section 5.4 it was shown that the optimal control problem was equivalent to maximizing an augmented index J^*

$$J^* = \phi(x(t_f), t_f) + \int_{t_0}^{t_f} \{H - p^{\mathrm{T}}\dot{x}\}\,dt$$
$$= \phi(x(t_f), t_f) + p(t_0)\,x(t_0) - p(t_f)\,x(t_f)$$
$$+ \int_{t_0}^{t_f} \{H + \dot{p}^{\mathrm{T}}x\}\,dt \tag{6.88}$$

where H is the Hamiltonian function

$$H(x, p, u, t) = p(t) f(x, u, t) + L(x, u, t)$$

and p is the adjoint variable.

Now consider a second-order perturbation in J^* about some nominal control u^0 and state x^0, i.e.,

$$\delta J^* = \delta^1 J^* + \tfrac{1}{2}\, \delta x^{\mathrm{T}} \phi_{xx}\, \delta x + \int_{t_0}^{t_f} \{\tfrac{1}{2}\, \delta x^{\mathrm{T}} H_{xx}\, \delta_x + \delta x^{\mathrm{T}} H_{xu}\, \delta u + \tfrac{1}{2}\, \delta u^{\mathrm{T}} H_{uu}\, \delta u\}\, dt \tag{6.89}$$

where

$$H_{xx} = p f_{xx} + L_{xx}, \qquad H_{xu} = p f_{xu} + L_{xu} \tag{6.90}$$

and

$$H_{uu} = p f_{uu} + L_{uu}$$

With the exception of the first term, the right-hand side of Equation (6.89) constitutes the second variation, i.e.,

$$\delta^2 J^* = \tfrac{1}{2}\, \delta x^{\mathrm{T}} \phi_{xx}\, \delta x + \int_{t_0}^{t_f} \{\tfrac{1}{2}\, \delta x^{\mathrm{T}} H_{xx}\, \delta x + \delta x^{\mathrm{T}}\, H_{xu}\, \delta u + \tfrac{1}{2}\, \delta u^{\mathrm{T}} H_{uu}\, \delta u\}\, dt \tag{6.91}$$

Clearly this equation is identical to Equation (6.87) because $p = V_x$.

6.6.1 The Convexity Condition

Now, for a maximum the second variation must be shown to be negative. This *convexity condition*, together with the stationarity condition, results in a general sufficiency theorem for *local* optimality.

The criteria needed to establish the negative definiteness of the second variation are provided by the successive sweep method. For, on the final iteration of the successive sweep algorithm a control law is produced which *maximizes the second variation*. The expression for the foregoing δu is given by

$$\delta u = -(V_x^0 f_{uu} + L_{uu})^{-1} (f_{ux} V_x^{0\mathrm{T}} + L_{ux} + f_u^{\mathrm{T}} V_{xx}^0)\, \delta x \tag{6.92}$$

which is identical to Equation (6.67).

But, if there are no deviations from the optimal, δu and hence δx is zero. Hence, the *upper bound* on the second variation is zero. Thus, the second variation must be negative definite for all perturbations so long as δu gives a true *unique* maximum in a local sense.

Hence, in order to show that $\delta^2 V < 0$, it is only necessary to show that the successive sweep solution exists and is unique. The criteria are

$$1. \; V^0_{xx} f_{uu} + L_{uu} < 0 \tag{6.93}$$

This is equivalent to the classical Legendre condition.

2. V^0_{xx} exists and is bounded. This is equivalent to Jacobi's condition or the conjugate point condition which is discussed further in Appendix A.

3. If control parameters are present,

$$V_{\alpha\alpha}(x(t_0), t_0) < 0 \tag{6.94}$$

4. If the terminal time is free, and no terminal constraints are present,

$$Z_{t_f t_f} < 0 \tag{6.95}$$

where $Z_{t_f t_f}$ is given by Equation (6.50).

Although no further proof is required to demonstrate that the foregoing conditions do guarantee that the second variation is negative definite, a simple direct proof will be given for the simple case without control parameters or side constraints.

Equation (6.87) may be rewritten

$$\begin{aligned}\delta^2 V = \tfrac{1}{2}\epsilon^2 \int_{t_0}^{t_f} \{&\delta\beta^{\mathrm{T}}(V_{txx} + V_{txxx} f + \dot{V}_{xx} f_x + f_x^{\mathrm{T}} V_{xx} + V_x f_{xx} + L_{xx})\,\delta\beta \\ &+ 2\delta\beta^{\mathrm{T}}(V_{xx} f_x + V_x f_{xu} + L_{xu})\,\delta\eta + \delta\eta^{\mathrm{T}}(V_x f_{uu} + L_{uu})\,\delta\eta\}\,dt\end{aligned} \tag{6.96}$$

Now from the characteristic equation for $\dot{V}_{xx}$, Equation (6.11),

$$\dot{V}_{xx} = V_{txx} + V_{xxx} f = -Z_{xx} + Z_{xu} Z_{uu}^{-1} Z_{xu}^{\mathrm{T}}$$

Thus, the first term under the integral, Equation (6.96), may be replaced by

$$\begin{aligned}&\delta\beta^{\mathrm{T}} Z_{xu} Z_{uu}^{-1} Z_{xu}^{\mathrm{T}}\,\delta\beta \\ &\quad = \delta\beta^{\mathrm{T}}(V_{xx} f_u + V_x f_{xu} + L_{ux})(V_x f_{uu} + L_{uu})^{-1}(V_{xx} f_u + V_x f_{xu} + L_{ux})\,\delta\beta\end{aligned}$$

The integral may now be written

$$\delta^2 V = \tfrac{1}{2}\epsilon^2 \int_{t_0}^{t_f} \|(V_{xx} f_u + V_x f_{xu} + L_{xu})^{\mathrm{T}}\,\delta\beta + (V_x f_{uu} + L_{uu})\,\delta\eta\,\|_Q^2\,dt$$

where $Q = (V_x f_{uu} + L_{uu})^{-1}$. Clearly $\delta^2 V$ is negative provided Q exists and is negative (condition 1).

PROBLEMS

1. Obtain the successive sweep algorithm from the discrete algorithm by using a finite difference technique.
2. Use the successive sweep algorithm to maximize the performance index J
$$J = -\int_0^1 (x_1^2 + u^2)\,dt$$
where $\dot{x}_1 = x_2 + u$, $x_1(0) = 1$, $\dot{x}_2 = -u$, and $x_2(0) = -1$. Show that convergence is obtained in one iteration from any priming solution.
3. Program the successive sweep algorithm and use it to duplicate the results of Examples 6.1 and 6.2.
4. One approach to the aircraft landing problem is to use *height*, as opposed to time, as the independent variable. What are the advantages of this approach? Specify how the successive sweep algorithm would be used to solve this problem.
5. Derive the boundary conditions for V_x and V_{xx} at the final time in the free-terminal time problem where a terminal constraint was employed as a stopping condition.

BIBLIOGRAPHY AND COMMENTS

For a different approach to the successive sweep algorithm, see

McReynolds, S. R. and Bryson, A. E. (1965). "A Successive Sweep Method for Solving Optimal Programming Problems," *Joint Autom. Control Conf.*, pp. 551–555 Troy, New York, June.

which in this form is very similar to an algorithm given by Mitter,

Mitter, S. K. (1966). "Successive Approximation Methods for the Solution of Optimal Control Problems," *Automatica*, Vol. 3, pp. 135–149.

A dynamic programming derivation is given in

McReynolds, S. R. (1967). "The Successive Sweep Method and Dynamic Programming," *J. Math. Anal. Appl.*, Vol. 19, No. 3, pp. 565–598.

For a discussion of constant feedback gains, see

Dyer, P. and McReynolds, S. R. (1969). "Optimization of Control Systems with Discontinuities and Terminal Constraints," *Trans. IEEE Autom. Control*, Vol. AC-14, pp. 223–229, June.

For several other similar algorithms, see

Jacobsen, D. H. (1968). "Second Order and Second Variation Methods for Determining Optimal Control. A Comparative Study Using Differential Dynamic Programming," *Intern. J. Control*, Vol. 7, No. 2, pp, 75–196.

Mayne, D. (1966). "A Second Order Gradient Method for Determining Optimal Trajectories of Non-Linear Discrete Time Systems," *Intern. J. Control*, Vol. 3, p. 85.

Merrian, C. W. (1964). *Optimization Theory and the Design of Feedback Control Systems.* McGraw-Hill, New York.

Noton, A. R. M., Dyer, P., and Markland, C. (1966). "Numerical Computation of Optimal Control," *Trans. IEEE Autom. Control*, Vol. AC. 12, No. 1, pp. 59–66.

Further detailed results on the aircraft and boiler examples may be found in

Dyer, P. (1966). "Application of Variational Methods to Aircraft Control," Ph.D. thesis. Univ. of Nottingham, England.

Markland, C. A. (1966). "Optimal Control of a Boiler," Ph.D. thesis. Univ. of Nottingham, England.

7

Systems with Discontinuities

7.1 INTRODUCTION

In the preceding chapters, it has been implicitly assumed that the return function, the performance index, and the state equations were sufficiently smooth so that the needed partial derivatives existed. One class of problems for which these assumptions are not valid is characterized by discontinuities in the derivatives of the state variable, that is to say, if τ denotes the time of a discontinuity

$$\dot{x}(\tau^-) \neq \dot{x}(\tau^+) \tag{7.1}$$

where $\dot{x}(\tau^-)$ is the left-hand limit of $\dot{x}$ at τ, i.e.,

$$\dot{x}(\tau^-) = \lim_{t \uparrow \tau} f(x(t), u(t), t) \tag{7.2}$$

and

$$\dot{x}(\tau^+) = \lim_{t \downarrow \tau} f(x(t), u(t), t) \tag{7.3}$$

One important type of problem in which these discontinuities occur is the *bang-bang* problem. In this type of problem the control variable takes on a finite set of values. For example, the optimal solution to a problem characterized by a linear system with a linear cost function and a bounded control will usually be of the bang-bang type. A practical example, which is considered in Section 7.3, is the minimum fuel control of the attitude of an orbiting body. In this problem the control has only three levels: maximum thrust in the positive or negative direction and zero thrust. The times at which the thrust changes are referred to as *switching* times. For any given sequence of control levels, the switching times determine the control history. Hence, in such cases, the switching times may be regarded as independent control parameters.

Although the theory developed in the previous chapters applies to the regions between such discontinuities, it does not apply in the neighborhood of a discontinuity. Further analysis is required to patch together the solutions from the regions on either side. In particular, in order to implement the gradient and successive sweep algorithms, the relationships between the partial derivatives of the return at either side of the discontinuity $V(x(\tau^+), \tau^+)$, and $V(x(\tau^-), \tau^-)$ must be found. Furthermore, an iterative technique must be found in order to compute the optimal switching times.

These problems with discontinuities may be conveniently divided into two classes. In one class the state is assumed to be continuous, whereas in the other, discontinuities in the state may occur. The first type of problem is considered in Sections 7.2–7.5, and the second in Sections 7.6–7.8.

7.2 DISCONTINUITIES: CONTINUOUS STATE VARIABLES

As discussed before the first class of problems is characterized by discontinuities in the first time derivative of the state variables, although the variables themselves are continuous. In other words, if τ denotes the time of a discontinuity, the following relations hold:

$$\dot{x}^- \neq \dot{x}^+ \qquad \text{and} \qquad x^- = x^+ \tag{7.4}$$

where the $-$ and $+$ refer to the left-hand and right-hand sides of the discontinuity. The performance index J

$$J = \phi(x(t_f), t_f) + \int_{t_0}^{t_f} L(x, u, \alpha, t)\, dt \tag{7.5}$$

is a continuous function of the initial time t_0 and so it follows that the associated return must also be continuous, i.e.,

$$V(x^-, \alpha, \nu, \tau^-) = V(x^+, \alpha, \nu, \tau^+) \tag{7.6}$$

7.2.1 Extension of the Gradient and Sweep Algorithms

The gradient and sweep methods must be re-examined in regions containing discontinuities if the discontinuities are displaced in time. In order to extend these algorithms, an expansion of the return function in terms of the switching time must be obtained. This expansion is required to join together the solutions obtained by the algorithms at either side of the discontinuity, and it is obtained by using the technique of strong variations. The expansion will be to second order about a single nominal switching time denoted by τ^j.

The extension to the gradient and successive sweep algorithm must be such that it enables a new nominal switching time, τ^{j+1}, $t_0 < \tau^{j+1} < t_f$, to be chosen to increase the value of the return. Let $V^-(x_0, t_0, \alpha, \nu; \tau^{j+1})$ denote the return function associated with the new switching time when the initial time and state are t_0 and x_0. This return V^- may be written

$$V^-(x_0, t_0, \alpha, \nu; \tau^{j+1}) = V^+(x(\tau^{j+1}), \tau^{j+1}, \alpha, \nu) + \int_{t_0}^{\tau^{j+1}} L(x, u^-, \alpha, t)\, dt \qquad (7.7)$$

where u^- is the control used to the left of the discontinuity and V^+ is the return function to the right of the discontinuity. The integral on the right-hand side of the foregoing equation is integrated along the trajectory x defined by

$$\dot{x} = f(x, u^-, \alpha, t), \qquad x(t_0) = x_0 \qquad (7.8)$$

To simplify the following analysis, the arguments of the various functions will be dropped unless they are needed to clarify the text. The + and − signs will be used as superscripts to distinguish between functions to the left and to the right of the discontinuity.

The right-hand side of Equation (7.7) is now expanded to second order about the nominal switching time τ^j, $x(\tau^j)$,

$$V^- = V^+(x(\tau^j), \tau^j, \alpha, \nu) + \int_{t_0}^{\tau^j} L^-\, dt + (L^- + V_t^+)\, d\tau + V_x^+\, dx$$
$$+ \tfrac{1}{2}[dx^{\mathrm{T}} V_{xx}^{\mathrm{T}}\, dx + 2V_{xt}^{+}\, dx\, d\tau + (\dot{L}^- + V_{tt}^{\mathrm{T}})\, d\tau^2] \qquad (7.9)$$

where $dx = x(\tau^{j+1}) - x(\tau^j)$, $d\tau = \tau^{j+1} - \tau^j$ and the partials of V^+ are evaluated at $x(\tau^j)$, τ^j.

Now, to second order,

$$dx = f^-\,|_{x^j\tau^j}\, d\tau + \tfrac{1}{2}\dot{f}^-\,|_{x^j\tau^j}\, d\tau^2 \qquad (7.10)$$

Substituting in Equation (7.9) for dx gives

$$V^- = V^+ + \int_{t_0}^{\tau^j} L^-\, dt + (L^- + V_t^+ + V_x^+ f^-)\, d\tau$$
$$+ \tfrac{1}{2}[V_x^+ \dot{f}^- + f^{-\mathrm{T}} V_{xx}^+ f^- + 2V_{tx}^+ f^- + \dot{L}^- + V_{tt}^+]\, d\tau^2 \qquad (7.11)$$

In order to evaluate this expression, the various partial derivatives of V^+ must be obtained. The partial derivatives with respect to x may be computed using their characteristic equations. The remaining time derivatives can be obtained directly from the dynamic programming equation for the return function, i.e.,

$$V_t^+ = -V_x^+ f^+ - L^+ \tag{7.12}$$

Differentiating this relationship with respect to x and t, in turn, yields

$$V_{tx}^+ = -V_{xx}^+ f^+ - V_x^+ f_x^+ - L_x^+ - (V_x^+ f_u + L_u)\, u_x^+ \tag{7.13}$$

and

$$V_{tt}^+ = -V_{tx}^+ f^+ - V_x^+ f_t^+ - L_t^+ - (V_x^+ f_u^+ + L_u^+)\, u_t^+ \tag{7.14}$$

The elimination of V_{tx}^+ from this expression for V_{tt}^+ results in

$$V_{tt}^+ = f^+ V_{xx}^+ f^+ - \dot{V}_x^+ f_x - V_x^+ f_t^+ - L_t^+ - (V_x^+ f_u^+ + L_u^+)\, \dot{u}^+ \tag{7.15}$$

Now substituting Equations (7.12), (7.13), and (7.15) for V_t^+, V_{tx}^+, V_{tt}^+ into Equation (7.11), one obtains

$$V^- = V^+ + \int_{t_0}^{\tau^j} L^-\, dt + V_\tau\, d\tau + \tfrac{1}{2} V_{\tau\tau}\, d\tau^2 \tag{7.16}$$

where

$$V_\tau = L^- - L^+ + V_x^+ (f^- - f^+) \tag{7.17}$$

and

$$\begin{aligned} V_{\tau\tau} = {} & (f^- - f^+)^{\mathrm{T}}\, V_{xx}^+ (f^- - f^+) - (V_x^+ f_x^+ + L_x^+)(f^- - f^+) \\ & + (V_x^+ (f_x^- - f_x^+) + L_x^- - L_x^+)\, f^- \\ & + V_x^+ (f_t^- - f_t^+) + L_t^- - L_t^+ - (V_x^+ f_u^+ + L_u^+)\, \dot{u}^+ \end{aligned} \tag{7.18}$$

Note that the last term in the expression for $V_{\tau\tau}$ will generally be zero for bang-bang problems.

Next, the effect of perturbations in the state x, control parameters α, and multipliers ν at the discontinuity will be investigated. First, in Equation (7.16) let t_0 approach τ^j so that the equation may be written

$$V^-(x^-(\tau^{j+1}), \nu, \alpha, \tau^{j+1}) = V^+(x^+(\tau^j), \nu, \alpha, \tau^j) + V_\tau\, d\tau + \tfrac{1}{2} V_{\tau\tau}\, d\tau^2 + \cdots \tag{7.19}$$

Expanding the right-hand side about the nominal $x^j(\tau^j)$, ν^j, and α^j, one obtains, to second order,

$$\begin{aligned} V^-(x^-(\tau^{j+1}), \nu, \alpha, \tau^{j+1}) \\ &= V^+(x^j(\tau^j), \nu^j, \alpha^j, \tau^j) + V_x^+\,\delta x + V_\tau\, d\tau + V_\nu^+\,\delta\nu + V_\alpha^+\,\delta\alpha \\ &\quad + \tfrac{1}{2}[\delta x^{\mathrm{T}} V_{xx}^+\,\delta x + \delta\alpha^{\mathrm{T}} V_{\alpha\alpha}^+\,\delta\alpha + \delta\nu^{\mathrm{T}} V_{\nu\nu}^+\,\delta\nu \\ &\quad + 2\delta x^{\mathrm{T}} V_{x\nu}^+\,\delta\nu + 2\delta x^{\mathrm{T}} V_{x\alpha}^+\,\delta\alpha + 2\delta\nu^{\mathrm{T}} V_{\nu\alpha}^+\,\delta\alpha \\ &\quad + 2\delta x^{\mathrm{T}} V_{x\tau}\, d\tau + 2\delta\nu^{\mathrm{T}} V_{\nu\tau}\, d\tau + 2\delta\alpha^{\mathrm{T}} V_{\alpha\tau}\, d\tau + V_{\tau\tau}\, d\tau^2] \end{aligned} \tag{7.20}$$

where

$$\begin{aligned} V_{x\tau} &= L_x^- - L_x^+ + L_u^- u_x^- - L_u^+ u_x^+ + V_{xx}^+(f^- - f^+) \\ &\quad + V_x^+(f_x^- - f_x^+ + f_u^- u_x^- - f_u^+ u_x^+) \\ V_{\alpha\tau} &= L_\alpha^- - L_\alpha^+ + L_u^- u_\alpha^- - L_u u_\alpha^+ + V_{x\alpha}^+(f^- - f^+) \\ &\quad + V_x^+(f_\alpha^- - f_\alpha^+ + f_u^- u_\alpha^- - f_u^+ u_\alpha^+) \\ V_{\nu\tau} &= V_{x\nu}^+(f^- - f^+) + (V_x^+ f_u^- + L_u^-)\, u_\nu^- - (V_x^+ f_u^+ + L_u^+)\, u_\nu^+ \end{aligned} \tag{7.21}$$

and

$$\delta x = x^{j+1} - x^j; \qquad \delta\alpha = \alpha^{j+1} - \alpha^j; \qquad \delta\nu = \nu^{j+1} - \nu^j \tag{7.22}$$

Note that the subscript τ in the foregoing partials denotes the *switching time*, not the initial time, and hence these partials are not to be confused with those given in Equation (7.13), etc.

Apparently the switching time is a new parameter that must be introduced into the return function. However, this is avoided by immediately eliminating $d\tau$, either by a first-order (gradient) or second-order (sweep) method.

7.2.2 The Gradient Algorithm

In deriving the gradient algorithm only the first-order terms in Equation (7.20) are considered, viz,

$$V^-(\tau^{j+1}) = V^+(x^j, \nu^j, \alpha^j, \tau^j) + V_x^+\,\delta x + V_\tau\, d\tau + V_\nu^+\,\delta\nu + V_\alpha^+\,\delta\alpha \tag{7.23}$$

Now $d\tau$ must be chosen to ensure that the new return $V^-(\tau^{j+1})$ is greater than the nominal return $V^-(\tau^j) = V^+(\tau^j)$. Clearly, $d\tau$ must be chosen so that

$$V_\tau\, d\tau > 0 \tag{7.24}$$

for example,

$$d\tau = \epsilon \operatorname{sgn}\{V_\tau\} \tag{7.25}$$

or

$$d\tau = \epsilon V_\tau \tag{7.26}$$

where ϵ is a step-length parameter chosen to ensure the validity of the expansion, Equation (7.23).

7.2.3 The Successive Sweep Algorithm

The second-order correction is obtained by choosing $d\tau$ to maximize the quadratic expansion on the right-hand side of Equation (7.20). The partial derivative of this equation with respect to $d\tau$ is

$$V_\tau + V_{\tau x}\,\delta x + V_{\tau\nu}\,\delta\nu + V_{\tau\alpha}\,\delta\alpha + V_{\tau\tau}\,d\tau = \partial V^- / \partial(d\tau)$$

If $V_{\tau\tau} < 0$, we may set the foregoing equation to zero and solve for $d\tau$, i.e.,

$$0 = V_\tau + V_{\tau x}\,\delta x + V_{\tau\nu}\,\delta\nu + V_{\tau\alpha}\,\delta\alpha + V_{\tau\tau}\,d\tau \tag{7.27}$$

and hence

$$d\tau = -V_{\tau\tau}^{-1}[V_\tau + V_{\tau x}\,\delta x + V_{\tau\nu}\,\delta\nu + V_{\tau\alpha}\,\delta\alpha] \tag{7.28}$$

However, this technique may be unsatisfactory if $V_{\tau\tau}$ is small (or zero) or positive.

Another approach, which may be used if terminal constraints are present, is based on the condition that $\delta\nu$ must be chosen to extremize the second-order expansion of V, i.e., $\delta\nu$ must be chosen such that

$$V_\nu^+ + V_{\nu x}^+\,\delta x + V_{\nu\alpha}^+\,\delta\alpha + V_{\nu\tau}\,d\tau + V_{\nu\nu}^+\,\delta\nu = 0 \tag{7.29}$$

The technique is to solve simultaneously the ith equation of Equation (7.29) together with Equation (7.27) for $\delta\nu_i$ and $\delta\tau$, i.e.,

$$\begin{bmatrix} \delta\tau \\ \delta\nu_i \end{bmatrix} = -\begin{bmatrix} V_{\nu_i\nu_i}^+ & V_{\nu_i\tau} \\ V_{\nu_i\tau} & V_{\tau\tau} \end{bmatrix}^{-1}$$

$$\times \left\{ \begin{bmatrix} V_\tau \\ V_{\nu_i}^+ \end{bmatrix} + \begin{bmatrix} V_{\tau x} \\ V_{\nu_i x}^+ \end{bmatrix} \delta x + \begin{bmatrix} V_{\tau\alpha} \\ V_{\nu_i\alpha}^+ \end{bmatrix} \delta\alpha + \begin{bmatrix} V_{\tau\nu'} \\ V_{\nu_i\nu'}^+ \end{bmatrix} \delta\nu' \right\} \tag{7.30}$$

where $\nu' = [\nu_i, \ldots, \nu_{i-1}, \nu_{i+1}, \ldots]^{\mathrm{T}}$.

This computation requires the inversion of a matrix with determinant D given by

$$D = V_{\nu_i\nu_i}^+ V_{\tau\tau} - V_{\nu_i\tau}^2 \tag{7.31}$$

In general, $V_{\nu_i\tau}$ should be chosen to be the largest element of the vector $V_{\nu\tau}$. An advantage of this procedure is that $\delta\nu_i$ is eliminated at the discontinuity thus reducing the number of multipliers ν.

Finally, $d\tau$ may be chosen to satisfy the equations (7.27) and (7.29) in a least squares sense, i.e.,

$$\begin{aligned} d\tau = &-[V_{\nu\tau}^{\mathrm{T}}V_{\nu\tau} + V_{\tau\tau}^2]^{-1} \\ &\times [V_\tau V_{\tau\tau} + V_\nu{}^+ V_{\nu\tau} + (V_{\tau\nu}V_{\nu x}^+ + V_{\tau\tau}V_{\tau x})\,\delta x \\ &+ (V_{\tau\nu}V_{\nu\nu}^+ + V_{\tau\tau}V_{\tau\nu})\,\delta\nu + (V_{\tau\alpha}V_{\alpha\alpha}^+ + V_{\tau\tau}V_{\tau\alpha})\,\delta\alpha] \end{aligned} \tag{7.32}$$

This requires the inversion of the scalar $(V_{\nu\tau}^{\mathrm{T}}V_{\nu\tau} + V_{\tau\tau}^2)$ which will almost always be nonzero.†

If none of the foregoing methods are suitable, a gradient method as discussed in the previous section should be employed.

7.2.4 Jump Conditions for the Partial Derivatives

All the foregoing methods eliminate $d\tau$ by means of a relationship of the form

$$d\tau = \tau_x\,\delta x + \tau_\alpha\,\delta\alpha + \tau_\nu\,\delta\nu + d\tau_0 \tag{7.33}$$

Substituting this expression for $d\tau$ back into the second-order expansion of the return function, Equation (7.20), results in a new expansion in the variables δx, $\delta\alpha$, and $d\nu$. By expanding the left-hand side and comparing coefficients of δx, $\delta\alpha$, etc., relationships are obtained between the partial derivatives of V^- and V^+, viz,

$$\begin{aligned} V_x{}^- &= V_x{}^+ + V_\tau\tau_x + d\tau_0 V_{\tau\tau}\tau_x + d\tau_0 V_{\tau x} \\ V_\nu{}^- &= V_\nu{}^+ + V_\tau\tau_\nu + d\tau_0 V_{\tau\tau}\tau_\nu + d\tau_0 V_{\tau\nu} \\ V_\alpha{}^- &= V_\alpha{}^+ + V_\tau\tau_\alpha + d\tau_0 V_{\tau\tau}\tau_\alpha + d\tau_0 V_{\tau\alpha} \\ V_{xx}^- &= V_{xx}^+ + V_{x\tau}\tau_x + \tau_x{}^{\mathrm{T}}V_{\tau x} + \tau_x{}^{\mathrm{T}}V_{\tau\tau}\tau_\nu \\ V_{x\nu}^- &= V_{x\nu}^+ + V_{x\tau}\tau_\nu + \tau_x{}^{\mathrm{T}}V_{\tau x} + \tau_x{}^{\mathrm{T}}V_{\tau\tau}\tau_\nu \\ V_{x\alpha}^- &= V_{x\alpha}^+ + V_{x\tau}\tau_\alpha + \tau_x{}^{\mathrm{T}}V_{\tau\alpha} + \tau_x{}^{\mathrm{T}}V_{\tau\tau}\tau_\alpha \\ V_{\alpha\alpha}^- &= V_{\alpha\alpha}^+ + V_{\alpha\tau}^+\tau_\alpha + \tau_\alpha{}^{\mathrm{T}}V_{\tau\alpha}^+ + \tau_\alpha{}^{\mathrm{T}}V_{\tau\tau}\tau_\alpha \\ V_{\alpha\nu}^- &= V_{\alpha\nu}^+ + V_{\alpha\tau}^+\tau_\nu + \tau_\alpha{}^{\mathrm{T}}V_{\tau\nu}^+ + \tau_\alpha{}^{\mathrm{T}}V_{\tau\tau}\tau_\nu \\ V_{\nu\nu}^- &= V_{\nu\nu}^+ + V_{\nu\tau}\tau_\nu + \tau_\nu{}^{\mathrm{T}}V_{\tau\nu} + \tau_\nu{}^{\mathrm{T}}V_{\tau\tau}\tau_\nu \end{aligned} \tag{7.34}$$

† If it is zero, $V_{\nu\tau}$ must be zero, which implies that the switching time has no effect, at least to first order, on the terminal constraints.

Thus, in general there will be a discontinuity in the partial derivatives of the return function at the discontinuity.

7.2.5 Neighboring Extremal Guidance

If the nominal trajectory is the optimal trajectory, some simplifications occur. The Newton–Raphson correction given in Equation (7.28) becomes, for the problem without control parameters or side constraints,

$$d\tau = -(V_{\tau\tau})^{-1} V_{\tau x} \delta x \tag{7.35}$$

This is a neighboring extremal guidance law for the switching time. It provides a linearized approximation to the switching surface. The equations relating the partial derivatives V_x and V_{xx} become

$$V_x^- = V_x^+, \qquad V_{xx}^- = V_{xx}^+ - V_{\tau x}^{\mathrm{T}} V_{\tau\tau}^{-1} V_{\tau x} \tag{7.36}$$

In other words, along the optimum trajectory V_x is continuous, although in general there will be a discontinuity in V_{xx}.

7.3 APPLICATION TO EXAMPLES

For the examples a combination of the gradient and successive sweep methods will be used. Whenever possible the successive sweep method will be used, but, if $V_{\tau\tau} > 0$, or, if the step length $V_\tau V_{\tau\tau}^{-1}$ is too large, a gradient correction will be made. The changes in the basic successive sweep procedure may be summarized as follows.

1. On the backward integration, the characteristic equations for V_x, V_{xx}, etc., are integrated backward to the first discontinuity. At the discontinuity the various functions V_τ, $V_{\tau\tau}$, etc., needed to form $d\tau$ are computed and stored. Then the derivatives are updated using Equation (7.34) and the backward integration proceeds.

2. At the next discontinuity step 1 is repeated and so on until the initial time is reached.

3. The state equations are integrated forward until the first switching time. The change $d\tau$ is computed using Equation (7.28), (7.30), or (7.32). If $d\tau$ is too large or if $V_{\tau\tau}$ is positive (for a maximum), a gradient step is taken.

EXAMPLE 7.1. *Minimum Fuel Control of an Orbiting Body.* In this example the problem of changing the overall attitude of an orbiting body with a minimum expenditure of fuel will be considered. It will be assumed that when the body reaches the desired terminal state some auxiliary control system will be used to ensure stability.

The equations describing the motion of a rigid body are quite well known. The basic equations may be obtained by choosing two three-axes cartesian systems, one of which is attached to the body and the other being inertially fixed. The origin of both systems is taken to be the center of mass. The angular motion of the body is described with respect to the body-fixed system. It remains to describe the orientation of the body with respect to the inertial system. There are a variety of ways with which this may be accomplished, for example, Euler angles, Cayley–Klein parameters, etc. Here a four parameter system, quaternions, will be used. Although one of these parameters is redundant, the equations do not become singular and are relatively simple in form.

The system equations are

$$\begin{aligned}
&\dot{x}_1 = u_1 - K_x x_2 x_3, \qquad \dot{x}_2 = u_2 - K_y x_1 x_3, \qquad \dot{x}_3 = u_3 - K_z x_1 x_2 \\
&\dot{x}_4 = (x_1 x_7 - x_2 x_6 + x_3 x_5)/2, \qquad \dot{x}_5 = (x_1 x_6 + x_2 x_7 - x_3 x_4)/2 \\
&\dot{x}_6 = (-x_1 x_5 + x_2 x_4 + x_3 x_7)/2, \qquad \dot{x}_7 = -(x_1 x_4 + x_2 x_5 + x_3 x_6)/2
\end{aligned} \tag{7.37}$$

The states x_1, x_2, and x_3 are angular velocities, and x_4, x_5, x_6 and x_7 are the variables describing position. (**Note:** As mentioned before, the parameters $x_4 - x_7$ are related, i.e., $x_4^2 + x_5^2 + x_6^2 + x_7^2 = 2$.) The control variables are denoted by u_1, u_2, and u_3, and the parameters K_x, K_y, K_z, which are functions of the moments of inertial of the body, are taken as -0.35125, 0.86058, and -0.73000, respectively.

The problem is to move from some initial state, here taken to be

$$\begin{aligned}
&x_1 = x_2 = x_3 = 1/57.3 \text{ rad/sec} \\
&x_4 = 0.4, \qquad x_5 = x_6 = 0.8
\end{aligned} \tag{7.38}$$

to a desired final state $x_f(t_f)$, here the origin, i.e.,

$$x_i(t_f) = 0, \qquad x_7 = 2, \qquad i = 1, 2, \ldots, 6 \tag{7.39}$$

while minimizing the fuel consumed. Thus the performance index is

$$J = \int_0^{t_f=60} \{|u_1| + |u_2| + |u_3|\}\, dt \tag{7.40}$$

The control u is constrained so that

$$|u_i| \leqslant 0.412/57.3 \text{ rad/sec}^2, \qquad i = 1, 2, 3 \tag{7.41}$$

The problem could be solved as posed with the hard terminal constraint. However, as the purpose of the control system is to correct gross errors in orientation or to make large changes in attitude, a penalty function approach will be used to ensure that the state is within an acceptable region.

Thus, the performance index becomes

$$J^* = \int_0^{60} \{| u_1 | + | u_2 | + | u_3 |\} \, dt + (x(60) - x_f)^{\mathrm{T}} \Lambda (x(60) - x_f) \tag{7.42}$$

The weighting matrix Λ is chosen to ensure that $[\sum_{i=1}^{6} x_i(60)^2]^{1/2} \leqslant 10^{-2}$ for the optimized trajectory. In this case, suitable values for the elements of the weighting matrix Λ were

$$\Lambda_{ii} = 0.5/(57.3)^2, \qquad i = 1, 2, 3$$

$$\Lambda_{ii} = 5, \qquad i = 4, 5, 6, 7$$

$$\Lambda_{ij} = 0, \qquad i \neq j$$

The problem now becomes that of minimizing J^*, Equation (7.42), while satisfying the state equations, Equations (7.37).

Now the optimal control must minimize [cf Equation (4.24)],

$$V_x^{\mathrm{opt}} f + L = \sum_{i=1}^{7} \{V_{x_i}^{\mathrm{opt}} f_i\} + | u_1 | + | u_2 | + | u_3 | \tag{7.43}$$

along the trajectory. It is easy to verify that u must be chosen so that

$$\left.\begin{aligned} u_i &= -\mathrm{sgn}\{V_{x_i}\} \mid V_{x_i}^{\mathrm{opt}} \mid \geqslant 1, \\ u_i &= 0 \mid V_{x_i}^{\mathrm{opt}} \mid < 1, \end{aligned}\right\} \quad i = 1, 2, 3 \tag{7.44}$$

i.e., u is of the bang-bang form. Thus, the analysis of the preceding section may be used to compute the optimal control.

The characteristic equations for V_x and V_{xx} needed for the successive sweep algorithm are

$$\dot{V}_x = -f_x{}^{\mathrm{T}} V_x, \qquad V_x(60) = 2\Lambda(x(60) - x_f) \tag{7.45}$$

and

$$\dot{V}_{xx} = -\{f_x{}^{\mathrm{T}} V_{xx} + V_{xx} f_x + V_x f_{xx}\}, \qquad V_{xx}(60) = 2\Lambda \tag{7.46}$$

where the matrices f_x and $V_x f_{xx}$ are

$$f_x = \begin{bmatrix} 0 & -K_x x_3 & -K_x x_2 & 0 & 0 & 0 & 0 \\ -K_y x_3 & 0 & -K_y x_1 & 0 & 0 & 0 & 0 \\ -K_z x_2 & -K_z x_1 & 0 & 0 & 0 & 0 & 0 \\ 0.5x_7 & -0.5x_6 & 0.5x_5 & 0 & 0.5x_3 & -0.5x_2 & 0.5x_1 \\ 0.5x_6 & 0.5x_7 & -0.5x_4 & -0.5x_3 & 0 & 0.5x_1 & 0.5x_2 \\ -0.5x_5 & 0.5x_4 & 0.5x_7 & 0.5x_2 & -0.5x_1 & 0 & 0.5x_3 \\ -0.5x_4 & -0.5x_5 & -0.5x_6 & -0.5x_1 & -0.5x_2 & -0.5x_3 & 0 \end{bmatrix}.$$

and

$$V_x f_{xx} = \begin{bmatrix} 0 & -\lambda_3 K_z & -\lambda_2 K_y & -0.5\lambda_7 & -0.5\lambda_6 & 0.5\lambda_5 & 0.5\lambda_4 \\ & 0 & -\lambda_1 K_x & 0.5\lambda_6 & -0.5\lambda_7 & -0.5\lambda_4 & 0.5\lambda_5 \\ & & 0 & -0.5\lambda_5 & 0.5\lambda_4 & -0.5\lambda_7 & 0.5\lambda_6 \\ & & & 0 & 0 & 0 & 0 \\ \text{symmetric} & & & & 0 & 0 & 0 \\ & & & & & 0 & 0 \\ & & & & & & 0 \end{bmatrix}$$

The additional relations needed at the switching times are

$$V_\tau = V_x^{\mathrm{T}}(\tau)\,\Delta f + |u_1^-| + |u_2^-| + |u_3^-| - \{|u_1^+| + |u_2^+| + |u_3^+|\}$$
$$V_{x\tau} = \Delta f^{\mathrm{T}} V_{xx}^{\mathrm{T}} \tag{7.47}$$

and

$$V_{\tau\tau} = \Delta f^{\mathrm{T}} V_{xx}^{\mathrm{T}} \Delta f - V_x^{+} f_x^{+} \Delta f \tag{7.48}$$

where

$$\Delta f = f^- - f^+ = [u_1^- - u_1^+, u_2^- - u_2^+, u_3^- - u_3^+, 0, 0, 0, 0]^{\mathrm{T}} \tag{7.49}$$

The equations (7.34), which are used to update the value of V_{xx}^+ and V_x^+ to V_{xx}^- and V_x^- at each switching time, complete the relations needed for the iterative procedure.

The computational procedure may now be summarized as follows:

1. Choose a suitable priming trajectory.
2. Integrate Equations (7.37) forward and store $\mathbf{x}$.
3. Integrate the differential equations (7.45) and (7.46) in reverse time until a switch is encountered.
4. Compute and store V_τ, $V_{x\tau}$, and $V_{\tau\tau}$. Compute V_x^- and V_{xx}^-.
5. Continue the integration to the next switch.
6. Repeat steps 4 and 5 until the initial time.

7. Integrate the state equations forward to the first switch. Compute $d\tau = -(V_{\tau\tau})^{-1}(V_{x\tau}\,\delta x + V_\tau)$ and form $\tau^{j+1} = \tau^j + d\tau$. (**Note**: At the first switch $\delta x = 0$ always).
8. Continue the integration storing the state x.
9. Compute the performance index.
10. Repeat steps 3–9 until no further improvement is made.
11. Check that the switching function is satisfied. Here

$$\begin{aligned} u_i &= -\text{sgn}(V_{x_i}), &&\text{for} \quad |V_{x_i}| \geqslant 1, \\ & && && i = 1, 2, 3 \\ u_i &= 0, &&\text{for} \quad |V_{x_i}| < 1, \end{aligned}$$

It should be noted that in general, several gradient steps would have to be taken before the full Newton–Raphson step could be used. In these cases, in step 7 $d\tau$ would be chosen as $d\tau = -\epsilon V_\tau$ and appropriate

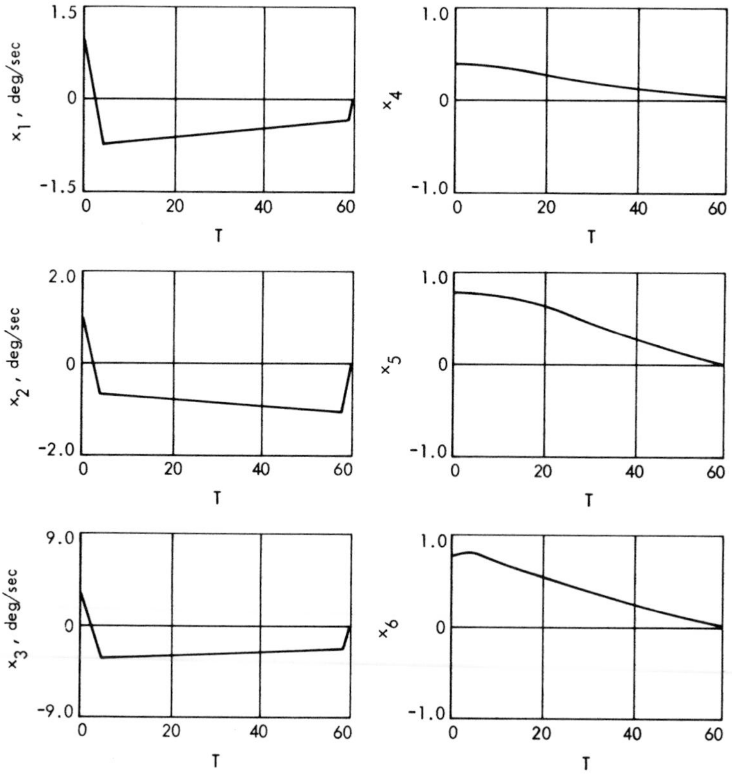

FIG. 7.1. Optimal solution.

modifications should be made to the equations used to form V_x^- and V_{xx}^-, i.e., set $\tau_x = 0$ in Equation (7.34).

One feature of this problem was that the priming trajectory had to be chosen carefully. One technique is to start with a series of narrow pulses. However, this was found to be impracticable as the system converged to an incorrect extremum. A more profitable approach, which was actually used to obtain the final solution, was to begin with a single impulse in each axis chosen so as to reduce the final position errors. These first control pulses were chosen near the initial time so that they would have the greatest effect on the trajectory. After these were optimized, a check of the switching function indicated that more switches were required in the vicinity of the final time. The nominal switching times were chosen to minimize errors in the terminal velocity. These times were then optimized.

The optimized trajectories are shown in Figures 7.1 and 7.2, and the

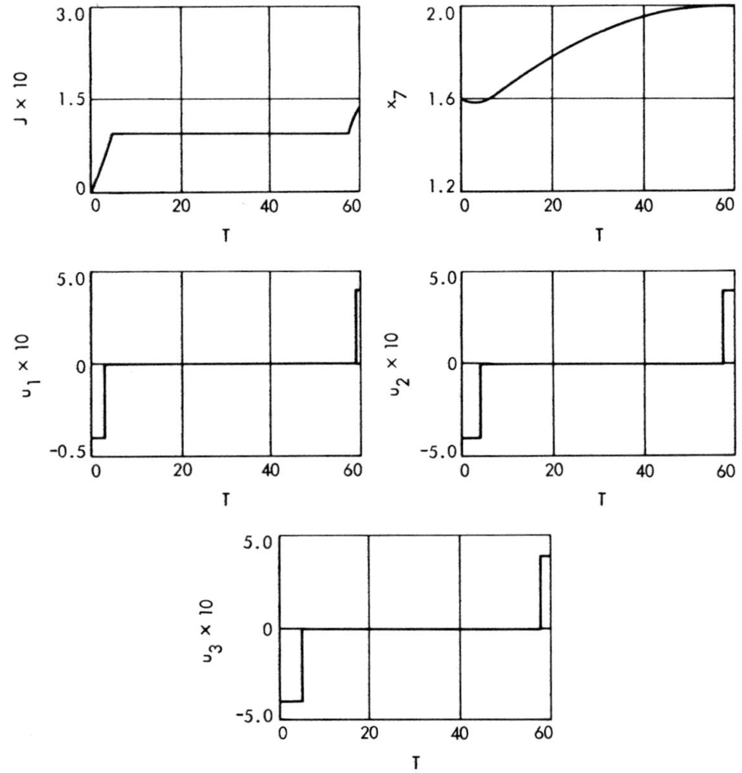

FIG. 7.2. Optimal solution.

convergence is shown in Table XXVII. As can be seen, seven steps were required even from a fairly good nominal solution. Note the very simple form of the final control, simply a pulse at each end of the trajectory. The control obtained using the successive sweep method satisfied all of the conditions for a minimum, and, hence, is at least locally optimal.

TABLE XXVII. CONVERGENCE OF THE SECOND VARIATIONS METHOD

Iteration Number	Cost	Switching Times					
		t_1	t_2	t_3	t_4	t_5	t_6
Prime	0.4101	3.500	4.000	5.000	57.500	58.000	59.000
1	0.2726	3.535[a]	4.035[a]	4.912[a]	57.625[a]	58.000[a]	59.127
2	0.2047	3.564[a]	4.073[a]	4.927	57.750	58.010	59.255
3	0.1612	3.606[a]	4.146[a]	4.962	57.618	58.009	59.382
4	0.1399	3.672[a]	4.145	5.011	57.562	58.003	59.446
5	0.1318	3.738	4.127	5.038	57.524	58.018	59.384
6	0.1303	3.780	4.116	5.055	57.499	58.028	59.344
7	0.1303	3.780	4.117	5.055	57.499	58.029	59.344
		Final values of $\ddot{V}$					
		1.635	1.254	5.376	0.854	0.854	0.849

[a] Gradient steps: $\Delta t = \epsilon V_T$.

A fixed integration step length of 0.5 sec was used with a Runge–Kutta integration scheme. The switching time was chosen within the limits of single precision arithmetic, although double precision was used in the integration scheme. One iteration (forward and backward integration) took about 23 sec on the IBM 7090 computer.

EXAMPLE 7.2. *Example* 7.1 *with Terminal Constraints.* As a further example of bang-bang control, the previous example is repeated except that the hard terminal constraint replaces the penalty function. First, the terminal constraint is adjoined to the performance criterion with multipliers ν_i, i.e.,

$$J^* = J + \sum_{i=1}^{6} \{\nu_i x_i(60)\} \tag{7.50}$$

As noted before, only six of the components of the state vector need be specified because the seventh state variable is a function of x_4, x_5, and x_6.

It is left to the reader to form the additional characteristic equations for V_ν, $V_{\nu\nu}$, and $V_{x\nu}$ and the partial derivative at the switching time $V_{\nu\tau}$.

The successive sweep algorithm was then used to compute the optimal control. The procedure was very similar to that used for the previous example except that $d\tau$ was chosen by Equation (7.32), viz,

$$d\tau = -[V_{\tau\nu}V_{\nu\tau} + V_{\tau\tau}^2]^{-1}[V_\tau V_{\tau\tau} + V_\nu V_{\nu\tau} + (V_{\tau\nu}V_{\nu x} + V_{\tau\tau}V_{\tau x})\,\delta x + (V_{\tau\nu}V_{\nu\nu} + V_{\tau\tau}V_{\tau\nu})\,\delta\nu] \tag{7.51}$$

TABLE XXVIII

CONVERGENCE OF THE SUCCESSIVE SWEEP METHOD: SWITCHING TIMES AND LAGRANGE MULTIPLIERS

Iteration Number	Fuel	Switching Times t_1	t_2	t_3	t_4	t_5	t_6
Prime	0.1294	3.500	4.000	5.000	57.500	58.000	59.000
1	0.1332	3.750	4.143	5.133	57.698	57.75	59.250
2	0.1289	3.794	4.118	5.053	57.690	58.000	59.348
3	0.1301	3.784	4.113	5.068	57.705	57.994	59.353
4	0.1310	3.798	4.129	5.072	57.454	58.026	59.302
5	0.1287	3.783	4.113	5.067	57.705	58.001	59.363
6	0.1306	3.791	4.121	5.070	57.476	58.013	59.330
7	0.1306	3.791	4.121	5.070	57.476	58.013	59.330

Iteration Number	Lagrange Multipliers 1	2	3	4	5	6
Prime	−1.08	−1.20	−1.11	0.03	0.046	0.029
1	−1.015	−1.119	−1.036	0.028	0.046	0.025
2	−1.020	−1.119	−1.055	0.030	0.045	0.028
3	−1.020	−1.130	−1.054	0.030	0.046	0.029
4	−1.020	−1.119	−1.054	0.030	0.046	0.029
5	−1.020	−1.131	−1.054	0.030	0.046	0.029
6	−1.020	−1.131	−1.054	0.030	0.046	0.029
7	−1.020	−1.131	−1.054	0.030	0.046	0.029

In this example the correction to ν was computed at the initial time. Note that a nominal value also had to be chosen for the multiplier ν. In this case advantage was taken of the previous penalty function solution and ν was chosen by

$$\nu_i = 2A_{ii}(x_i(60)), \qquad i = 1, 2, \ldots, 6$$

This is an example of a technique that can always be used and usually provides a very good approximation. In general, if the penalty function is of the form $P = (x - x_f)^{\mathrm{T}} \Lambda (x - x_f)$ the Lagrange multiplier should be chosen as $\nu = 2\Lambda(x^* - x_f)$ where x^* is the value of $x(t_f)$ obtained from the penalty function solution.

The optimal trajectories were very similar to those for the soft constraint case (Figure 7.1). Details of the convergence are shown in Table XXVIII. As can be seen, the nominal multipliers, obtained as described before, were extremely close to the optimal multipliers. Convergence was again fairly slow. It must be borne in mind that discontinuities represent very severe nonlinearities. The terminal constraints were met to at least five significant figures and probably higher accuracy could have been obtained by refining the computer program and by increasing the number of iterations. The minimum fuel was 0.1306, as compared to 0.13 for the soft terminal constraints. (The state trajectories were very similar to those for Example 7.1, Figure 7.1.)

Again, a fixed integration step of 0.5 sec was used with the Runge–Kutta integration scheme. One iteration took about 36 sec on the IBM 7090 computer.

EXAMPLE 7.3. *Lawden's Problem.* Next, consider the minimum-fuel orbit transfer of a low thrust rocket in a gravitational field. For the purpose of this problem the rocket is constrained to a plane. The rocket position may be characterized by a set of inertially fixed cartesian coordinates with the origin at the center of the gravitational field. The velocity is expressed in terms of its components in the cartesian directions. Control is provided by the direction θ and magnitude α of the thrust. Figure 7.3 illustrates this coordinate system.

In the state vector notation the system equations are given by

$$\begin{aligned} \dot{x}_1 &= x_3, \qquad \dot{x}_2 = x_4 \\ \dot{x}_3 &= -\mu x_1/(x_1^2 + x_2^2)^{3/2} + a\cos\theta/x_5 \\ \dot{x}_4 &= -\mu x_2/(x_1^2 + x_2^2)^{3/2} + a\sin\theta/x_5 \\ \dot{x}_5 &= -a/c \end{aligned} \tag{7.52}$$

where x_1 and x_2 denote the position, x_3 and x_4 denote the velocity, and x_5 denotes the mass.

The thrust a is constrained so that $0 \leqslant a \leqslant 0.1405$ and μ and c are system constants which are chosen as follows: $\mu = 1$; $c = 2$.

The initial state is chosen to be

$$x_1(0) = 1, \qquad x_2(0) = 0, \qquad x_3(0) = 0, \qquad x_4(0) = 1, \qquad x_5(0) = 0 \quad (7.53)$$

The terminal boundary conditions are

$$x_1(t_f) = 0, \qquad y_2(t_f) = 1.2 \tag{7.54}$$

and the final time t_f is unspecified. The problem is to maximize the final mass $m(t_f)$ of the rocket, i.e., to maximize the performance index J, where $J = x_5(t_f)$. The optimal control $u^{\text{opt}} = [\theta^{\text{opt}}, a^{\text{opt}}]^{\text{T}}$ is given by [cf. Equation (4.24)]

$$\theta^{\text{opt}}, a^{\text{opt}} = \underset{\theta, a}{\arg\max}\{V_{x_3}^{\text{opt}}\, a \cos\theta/x_5 - V_{x_5}^{\text{opt}}[a/c]\} \tag{7.55}$$

Thus it may be seen that the magnitude of the thrust a^{opt} is given by $a = 0.1405$ where $R > 0$ and $u = 0$ where $R < 0$ and where R is given by

$$R = (V_{x_3}^{\text{opt}} \cos\theta^{\text{opt}} + V_{x_4}^{\text{opt}} \sin\theta^{\text{opt}})/x_5 - V_{x_5}^{\text{opt}}/c$$

Clearly, as R changes sign there will be a discontinuity in the control. Actually, in this case the optimal solution consists of two parts, an initial period of variable thrust followed by a coast (zero thrust) to the final time.

As usual, the constraints are adjoined to the performance index to form a modified index J^*,

$$J^* = x_5(t_f) + \nu_1 x_1(t_f) + \nu_2(x_2(t_f) - 1.2)$$

where ν_1 and ν_2 are the Lagrange multipliers.

Now the successive sweep procedure could be applied in a straightforward manner. However, because of the structure of the problem, certain analytic simplifications may be made. First, consider a second-order expansion of the return at the nominal final time t_f [cf Equation (6.48)]

$$\begin{aligned}\delta V(t_f) = {} & [\nu_1 \nu_2 1][\delta x_1\, \delta x_2\, \delta x_5]^{\text{T}} + [x_1, (x_2 - 1.2)][\delta\nu_1, \delta\nu_2]^{\text{T}} \\ & + [\nu_1 x_3 + \nu_2 x_4 - a/c]\, dt_f + \delta x_1\, \delta\nu_1 + \delta x_2\, \delta\nu_2 \\ & + [x_3\, \delta\nu_1 + \nu_1\, \delta x_3 + x_4\, \delta\nu_2 + \nu_2\, \delta x_4]\, dt_f + \tfrac{1}{2} z_{t_f t_f}\, dt_f^2 \end{aligned} \tag{7.56}$$

where $z_{t_f t_f}$ is given by

$$z_{t_f t_f} = \nu_1 \frac{-x_3 x_1}{(x_1^2 + x_2^2)^{3/2}} + \frac{a \sin\theta}{x_5} + \nu_2 \left(\frac{-x_3 x_2}{(x_1^2 + x_2^2)^{3/2}}\right) + \frac{a \sin\theta}{x_5} \tag{7.57}$$

The first simplification results from using the terminal condition $x_1(t_f) = 0$ as a stopping condition. The associated multiplier is chosen to ensure that [cf Equation (5.83)]

$$\nu_1 x_3 + \nu_2 x_4 - a/c = 0 \tag{7.58}$$

i.e., ν_1 is chosen by

$$\nu_1 = \frac{-(\nu_2 x_4 - a/c)}{x_3} \tag{7.59}$$

Next, variations $d\nu_1$ and dt_f will be eliminated together [cf Equation (7.30)] by choosing $\delta\nu$ and dt_f to maximize the expansion. Differentiating the expansion with respect to dt_f and $\delta\nu_1$ and setting to zero gives

$$x_1 + \delta x_1 + x_3\, dt_f = 0 \tag{7.60}$$

$$\nu_1\, \delta x_3 + \nu_2\, \delta x_4 + x_3\, \delta\nu_1 + x_4\, \delta\nu_2 + z_{t_f t_f}\, dt_f = 0 \tag{7.61}$$

These equations may now be solved simultaneously for dt_f and $\delta\nu_1$. The values thus obtained are substituted back into Equation (7.56) to eliminate dt_f and $\delta\nu_1$ from the expansion. The reader should verify that the boundary conditions for the backward integration of the characteristic equations then become

$$\begin{bmatrix} V_{x_1}(\tau) \\ V_{x_2}(\tau) \\ V_{x_3}(\tau) \\ V_{x_4}(\tau) \\ V_{\nu_2}(\tau) \end{bmatrix} = \begin{bmatrix} \nu_1 + \dot{x}_1 z^*_{t_f t_f}/x_3^2 \\ \nu_2 \\ x_1\nu_1/x_3 \\ 1 \\ x_2 - 1.2 - x_1\nu_2/x_3 \end{bmatrix} \tag{7.62}$$

The nonzero elements of the second partials at the final time are

$$V_{x_1x_1} = z_{t_f t_f}/x_3^2, \qquad V_{x_1x_3} = \nu_1/x_3, \qquad V_{x_1x_4} = -\nu_2/x_3 \tag{7.63}$$

and $V_{x_4\nu_2} = 1$.

The derivation of the characteristic equations and the partial derivatives of V at the switching time is straightforward and is left to the reader.

A detailed outline of the algorithm is as follows:

1. From a nominal control variable obtain a nominal trajectory. The initial thrust angle was chosen to be 0.5 over the interval [0, 1.5]. From then on a coast (no thrust) was assumed. The final time was determined in order that $x(t_f) = 0$ was most nearly satisfied.

2. Obtain the partial derivatives of the return function at the final time. The nominal value for ν_2 was chosen to be 0.9 and ν_1 was determined from Equation (7.59).

3. Eliminate variations dt_f and $\delta\nu_1$ from the return function at the final time.

4. Integrate the characteristic equations backward to the switching time τ.

5. Compute partial derivatives of the return with respect to the switching time. Eliminate $d\tau$ and $d\nu_2$ [cf Equation (7.30)].

6. Integrate the characteristic equations of the partial derivatives back to the initial time. Compute and store the feedback gains. In this problem, piecewise-constant feedback gains were used. At the final time the first fixed gains were chosen as the optimal gains. The gains were updated when

$$\sum_{1} | k_i - u_{x_i} | \geqslant 1 \tag{7.64}$$

where k is the constant gain and u_x is the optimal gain.

7. Integrate the system equations forward using the feedback law to generate a new control function.

8. At the switching time obtain corrections to τ and ν_2.

9. Integrate the system equations forward to the final time. At the final time obtain corrections to t_f. Go to step 2 and repeat if necessary.

Several nominal values were tried for ν_2. Convergence was not particularly sensitive to the initial choice of this parameter. Corrections to the thrusting angle θ were bounded by 0.2. If the Newton–Raphson correction exceeded this bound, a gradient method was employed.

In order that the switching time be an effective way to bring about changes in $x_2(t_f)$, the sensitivity of $x_2(t_f)$ with respect to ν_2 should be large. On the optimal $V_{\nu_2\tau} = 0.1263$, and this was an effective method. However, when the specified value of $x(t_f)$ was increased, $V_{\nu_2\tau}$ was significantly decreased and this technique became ineffective.

The optimal solution is shown in Figure 7.3. Following an initial thrust period, there is a coast to the final time. The gains for the optimal trajectory are given in Table XXIX. Various other quantities associated with the optimal solution are listed in Table XXX.

TABLE XXIX. FEEDBACK GAINS

Time Interval	$10^2 \times \theta_x$	θ_y	$10^3 \times \theta_u$	θ_v	$10^5 \times \theta_m$
[0, 0.075]	−1.194	−2.847	4.123	0.8006	1.312
[0.075, 0.225]	−3.069	−1.869	−8.975	0.8193	1.235
[0.225, 0.450]	−4.928	−0.5785	−20.03	0.9587	1.192
[0.450, 0.675]	−5.510	0.3229	−19.03	1.283	1.207
[0.675, 0.906]	−4.265	0.8154	−0.0213	1.928	1.135

TABLE XXX. THE OPTIMAL SOLUTION FOR LAWDEN'S PROBLEM

Time	θ (rad)	x_1	x_2	x_3	x_4
0	1.360	1.000	0.000	0.000	1.000
0.075	1.371	0.997	0.075	−0.073	1.008
0.150	1.380	0.989	0.151	−0.143	1.009
0.225	1.387	0.975	0.227	−0.217	1.006
0.300	1.391	0.957	0.302	−0.287	0.997
0.375	1.394	0.933	0.376	−0.355	0.982
0.450	1.395	0.904	0.449	−0.420	0.963
0.525	1.394	0.870	0.520	−0.483	0.938
0.600	1.393	0.831	0.590	−0.542	0.909
0.675	1.390	0.788	0.656	−0.597	0.876
0.750	1.386	0.742	0.721	−0.647	0.840
0.850	1.380	0.674	0.802	−0.707	0.787
0.906	1.380	0.634	0.845	−0.737	0.756

Feedback Gains for Switching Time

$\tau_{x_1} = -1.664$	$\tau_{x_2} = -10.76$	$\tau_{x_3} = -1.287$
$\tau_{x_4} = -6.540$	$\tau_{x_5} = 0$	

Lagrange Multipliers

$\nu_1 = 0.1281$	$\nu_2 = 0.5561$

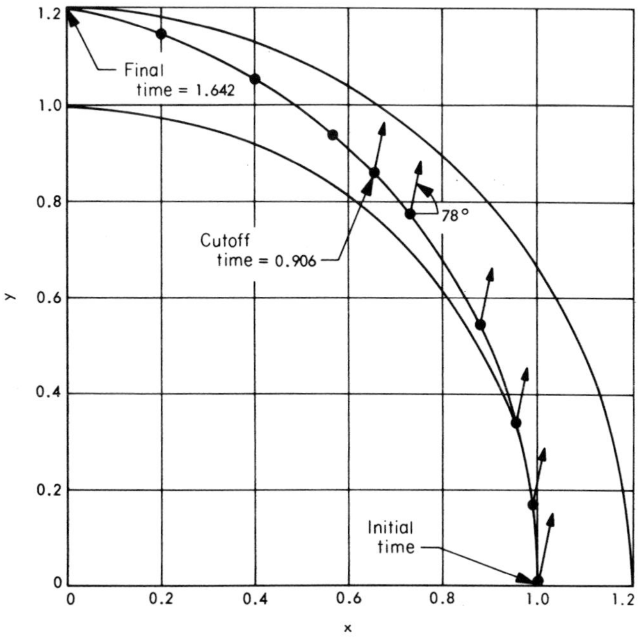

FIG. 7.3. Optimal low-thrust transfer.

7.4 THE FIRST VARIATION: A STATIONARITY CONDITION

The derivation of the first and second variations of the return with respect to the time of the discontinuity closely follows the development of Section 7.2.1.

Let the return associated with a nominal switching time τ^0 be denoted by $V(x_0, t_0; \tau)$. Now consider the return $V(x_0, t_0; \tau^1)$ associated with the switching time τ^1 where τ^1 is in the neighborhood of τ^0, i.e.,

$$\tau^1 = \tau^0 + \epsilon \tag{7.65}$$

The variation ϵ is chosen to be as small as desired.

Expanding the return $V(x_0, t_0; \tau^1)$ to first order about τ^0 one obtains (cf Section 7.2.1)

$$V(\tau^1) = V(\tau^0) + V_\tau \epsilon + 0(\epsilon^2) \tag{7.66}$$

where [cf Equation (7.17)] V_τ is given by

$$V_\tau = L^- - L^+ + V_x^+(f^- - f^+)|_{t=\tau^0} \tag{7.67}$$

Clearly, for τ^0 to be an extremal solution, the first variation, $V_\tau \epsilon$, must be nonpositive for arbitrary variations in ϵ. This will only be true provided the condition

$$V_\tau = 0 \tag{7.68}$$

holds. This is the *stationarity condition* which must hold for a local extremal solution provided no constraints are placed on the variations in τ.

7.5 THE SECOND VARIATION: A CONVEXITY CONDITION

Following the analysis of the previous section it can be shown that the expansion of the return $V(x_0, t_0; \tau^1)$ to second order is of the form

$$V(\tau^1) = V(\tau^0) + \epsilon V_\tau + \tfrac{1}{2}\epsilon^2 V_{\tau\tau} + O(\epsilon^3) \tag{7.69}$$

where V_τ is given by Equation (7.67) and [cf Equation (7.18)] $V_{\tau\tau}$ is given by

$$\begin{aligned} V_{\tau\tau} = {} & (f^- - f^+)^{\mathrm{T}} V_{xx}(f^- - f^+) - (V_x^+ f_x^+ + L_x^+)(f^- - f^+) \\ & + (V_x^+(f_x^- - f_x^+) + L_x^- - L_x^+)\, f^- + V_x^+(f_t^- - f_t^+) \\ & + L_t^- - L_t^+ - (V_x^+ f_u^+ + L_u^+)\, \dot{u}^+ \end{aligned} \tag{7.70}$$

For a maximum, the second variation, $\frac{1}{2}\epsilon^2 V_{\tau\tau}$, must be nonpositive, i.e.,

$$V_{\tau\tau} \leqslant 0 \tag{7.71}$$

This condition equation (7.71) is the *convexity* condition associated with a variable switching time.

EXAMPLE 7.4. *An Analytic Example.* An example will be given to illustrate the necessity of the foregoing analysis. In particular, it will be shown that a misleading result can be obtained by a straightforward solving of the dynamic programming equations.

Consider the problem of maximizing the performance index J where

$$J = \tfrac{1}{2}x_1^2(2) + \tfrac{1}{2}x_2^2(2)$$

with the system equations,

$$\dot{x}_1 = x_2 + u, \qquad x_1(O) = 0$$

$$\dot{x}_2 = -u, \qquad x_2(O) = 0$$

and the inequality constraint $|u| \leqslant 1$.

The candidate for the optimal solution is

$$u^0 = \begin{matrix} -1, & 0 \leqslant t \leqslant 1 \\ +1, & 1 \leqslant t \leqslant 2 \end{matrix}$$

It is easy to verify that this control maximizes the dynamic programming equation $(V_{x_1} - V_{x_2})\,u + x_2 V_{x_1}$. For the equation is a maximum when $u = \operatorname{sgn}(V_{x_1} - V_{x_2})$. Now V_{x_1} and V_{x_2} are given by

$$\begin{bmatrix} \dot{V}_{x_1} \\ \dot{V}_{x_2} \end{bmatrix} = -\begin{bmatrix} 0 & 0 \\ 1 & 0 \end{bmatrix} \begin{bmatrix} V_{x_1} \\ V_{x_2} \end{bmatrix}$$

with $V_{x_1}(2) = x_1(2) = 1$ and $V_{x_2}(2) = x_2(2) = 0$. Hence, $V_{x_1}(t) = 1$, $V_{x_2}(t) = 2 - t$, and $V_{x_1} - V_{x_2} = -1 + t$. Thus the switch is at $t = 1$ and, as for $t > 1$, $V_{x_1} - V_{x_2} > 0$, u^+ must be equal to $+1$. Thus the nominal satisfies and maximizes the dynamic programming equation.

Now the new condition, Equation (7.71), will be applied. The matrix of second partials V_{xx} is given by

$$\dot{V}_{xx} = -\begin{bmatrix} 0 & 0 \\ 1 & 0 \end{bmatrix} V_{xx} - V_{xx} \begin{bmatrix} 0 & 1 \\ 0 & 0 \end{bmatrix}$$

with

$$V_{xx}(2) = \begin{bmatrix} 1 & 0 \\ 0 & 1 \end{bmatrix}$$

Hence,

$$V_{xx}(t) = \begin{bmatrix} 1 & 2 - t \\ 2 - t & 1 + (2 - t)^2 \end{bmatrix}$$

also,

$$f^- - f^+ = 2\begin{bmatrix} -1 \\ +1 \end{bmatrix}$$

and

$$V_x{}^+ f_x{}^+ + L_x{}^+ - (V_x{}^- f_x{}^- + L_x{}^-) = 0$$

$$V_x{}^+ f_t{}^+ + L_t{}^+ - (V_x{}^- f_t{}^- + L_t{}^-) = 0$$

Thus $V_{\tau\tau}$ becomes

$$V_{\tau\tau} = (f^- - f^+)^{\mathrm{T}}\, V_{xx}(1)(f^- - f^+) - (V_x f_x)(f^- - f^+) = 2$$

As $V_{\tau\tau}$ is positive, the necessary condition is not satisfied and the proposed solution is not optimal. This result may be confirmed by evaluation of the performance index. For the nominal switching time t^*, J is given by

$$J = \tfrac{1}{2}[-(2 - 2t^*) - t^{*2} + (2 - 2t^*)^2]^2 + \tfrac{1}{2}[2 - 2t^*]^2$$

and substituting $t^* = 1$ gives $J = 0.5$. Now considering a small perturbation in t^*, $\Delta t = \pm 0.05$ gives

$$\begin{aligned} J &= \tfrac{1}{2}\left[\pm 0.1 - \begin{pmatrix}1.05\\0.95\end{pmatrix}^2 + 0.005\right]^2 + \tfrac{1}{2}[0.1]^2 \\ &= \tfrac{1}{2}[0.9975]^2 + \tfrac{1}{2}[0.1]^2 \\ &= 0.502603125 \end{aligned}$$

Thus the nominal solution is not optimal, as small variations in the switching time increase the cost.

In both the numerical examples the convexity condition given by Equation (7.71) was satisfied, and so the solutions obtained were at least locally optimal.

7.6 DISCONTINUITIES IN THE STATE VARIABLES

In the second class of problems discussed in Section 7.1 discontinuities may appear in both the control *and in the state vectors.* Such discontinuities appear, for example, in impulsive control problems such as the midcourse guidance of a space probe or the orbital trim of a planetary orbiter.

It will be assumed that the state at the right-hand side of the discontinuity x^+ is related to the state at the left-hand side x^- by an expression of the form

$$x^+(\tau) = g(x^-(\tau), \theta, \tau) \tag{7.72}$$

where θ is a vector of control parameters associated with the discontinuity, e.g., the magnitude and direction of the impulse. If some penalty is associated with this discontinuity, say $P(\theta, \tau)$, the return function is no longer continuous. The return at either side of the discontinuity is now related by

$$V^-(x^-(\tau), \alpha, \theta, \tau^-) = V^+(x^+(\tau), \alpha, \tau^+) + P(\theta, \tau) \tag{7.73}$$

7.6.1 Extension of the Gradient and Sweep Algorithms

The extension of the numerical algorithms follows closely the development of 7.2.1. Let $V^-(x_0, t_0, \tau^{j+1}, \theta^{j+1})$ denote the return function associated with the new switching time τ^{j+1} and parameters θ^{j+1} when the initial time and state are t_0 and x_0. This return V^- may be written

$$V^-(x_0, t_0, \tau^{j+1}, \theta^{j+1}) = V^+(x^+(\tau^{j+1}), \tau^{j+1}) + \int_{t_0}^{\tau^{j+1}} L(x, u^-, t)\, dt + P(\theta^{j+1}, \tau^{j+1}) \tag{7.74}$$

where u^- is the control used to the left of the discontinuity and V^+ is the return function to the right of the discontinuity. As before the arguments of the various functions will be dropped unless they are needed to clarify the text.

Equation (7.72) must now be used to eliminate x^+ from the first term of the right-hand side of Equation (7.74). However, to simplify the analysis further it will be assumed that $g(x^-, \theta)$ is given by

$$g(x^-, \theta) = x^- + h(\theta) \tag{7.75}$$

It is left to the reader to carry out the following analysis in the general case. Carrying out this substitution gives

$$V^-(x_0, \tau^{j+1}, \theta^{j+1}) = V^+(x^- + h(\theta^{j+1})) + \int_{t_0}^{\tau^{j+1}} L(x_1 u^-, t)\, dt + P(\theta^{j+1}, \tau^{j+1}) \tag{7.76}$$

Expanding the right-hand side of this equation to second order about the nominal switching time τ^j and nominal control parameter θ^j gives

$$\begin{aligned} V^-(x_0, \tau^{j+1}, \theta^{j+1}) = {} & V^+(x^- + h(\theta^j), \tau^j) + \int_{t_0}^{\tau^j} L(x, u^-, t)\, dt + P(\theta^j, \tau^j) \\ & + [V_\tau + P_\tau]\, d\tau + [V_x{}^+ h_\theta + P_\theta]\, \delta\theta \\ & + \tfrac{1}{2}[V_{\tau\tau} + P_{\tau\tau}]\, d\tau^2 + \tfrac{1}{2}\, \delta\theta^{\mathrm{T}}[h_\theta{}^{\mathrm{T}} V_{xx} h_\theta + V_x h_{\theta\theta} + P_{\theta\theta}]\, \delta\theta \\ & + [V_{\tau x} h_\theta + P_{\tau\theta}]\, d\tau\, \delta\theta \end{aligned} \tag{7.77}$$

where V_τ and $V_{\tau\tau}$ are given by Equations (7.17) and (7.18). The term $V_{\tau x}$ is given by Equation (7.21).

Next the effects of perturbations in the state x^- at the discontinuity will be examined. (It is left to the reader to extend the analysis to include variations in control parameters α and multipliers ν.) First let t_0 approach τ^j so that Equation (7.77) may be written

$$\begin{aligned} V^-(x^-, \tau^{j+1}, \theta^{j+1}) = {} & V^+(x^- + h(\theta^j), \tau^j) + Z_\tau\, d\tau + Z_\theta\, \delta\theta \\ & + \tfrac{1}{2}[Z_{\tau\tau}\, d\tau^2 + 2Z_{\tau\theta}\, d\tau\, \delta\theta + \delta\theta^{\mathrm{T}} Z_{\theta\theta}\, \delta\theta] \end{aligned} \tag{7.78}$$

where

$$\begin{aligned} Z_\tau &= V_\tau + P_\tau, & Z_\theta &= V_x{}^+ h_\theta + P_\theta \\ Z_{\tau\theta} &= V_{\tau x}^+ h_\theta + P_{\tau\theta}, & Z_{\tau\tau} &= V_{\tau\tau}^+ + P_{\tau\tau} \\ Z_{\theta\theta} &= h_\theta{}^{\mathrm{T}} V_{xx} h_\theta + V_x h_{\theta\theta} + P_{\theta\theta} \end{aligned} \tag{7.79}$$

Expanding the right-hand side of Equation (7.78) to second order gives

$$\begin{aligned} V(x^-, \tau^{j+1}, \theta^{j+1}) = {} & V^+(x^- + h(\theta^j), \tau^j) + Z_\tau \, d\tau + Z_\theta \, \delta\theta + V_x{}^+ \, \delta x^- \\ & + \tfrac{1}{2}[\delta x^{\mathrm{T}} V_{xx}^+ \, \delta x + 2\delta x^{\mathrm{T}} V_{x\tau} \, d\tau + 2\delta x^{\mathrm{T}} Z_{x\theta} \, \delta\theta \\ & + 2Z_{\tau\theta} \, d\tau \, \delta\theta + \delta\theta^{\mathrm{T}} Z_{\theta\theta} \, \delta\theta + Z_{\tau\tau} \, d\tau^2] \end{aligned} \tag{7.80}$$

where $Z_{x\theta}$ is given by $Z_{x\theta} = V_{xx}^+ h_\theta$ and, of course, $\delta x = x^{j+1} - x^j$, $\delta\theta = \theta^{j+1} - \theta^j$, and $d\tau = \tau^{j+1} - \tau^j$.

Again the switching time and the control parameters θ are new parameters that must apparently be introduced into the return function. However, this is avoided by immediately eliminating $d\tau$ and $\delta\theta$, either by a first-order (gradient) or a second-order (sweep) method.

7.6.2 The Gradient Algorithm

In deriving the extension to the basic gradient algorithm only the first-order terms in Equation (7.20) are considered, viz,

$$V^-(x^-, \theta^{j+1}, \tau^{j+1}) = V^+(x^- + h(\theta^j), \tau^j) + Z_\tau \, d\tau + Z_\theta \, \delta\theta + V_x{}^+ \, \delta x^- \tag{7.81}$$

Now $d\tau$ and $\delta\theta$ must be chosen to ensure that the new return $V^-(x^-, \tau^{j+1})$ is greater than the nominal return. Clearly $d\tau$ must be chosen so that $V_\tau \, d\tau > 0$ and $Z_\theta \, \delta\theta > 0$, for example,

$$d\tau = \epsilon V_\tau , \qquad \delta\theta = \epsilon Z_\theta \tag{7.82}$$

where ϵ, a step-length parameter, is chosen to ensure the validity of the expansions.

7.6.3 The Successive Sweep Algorithm

The second-order correction is obtained by choosing $d\tau$ and $\delta\theta$ to maximize the quadratic expansion on the right-hand side of Equation (7.80), i.e., $d\tau$ and $\delta\theta$ are chosen by

$$\begin{bmatrix} d\tau \\ \delta\theta \end{bmatrix} = - \begin{bmatrix} Z_{\tau\tau} & Z_{\tau\theta} \\ Z_{\theta\tau} & Z_{\theta\theta} \end{bmatrix}^{-1} \left[\begin{bmatrix} Z_\tau \\ Z_\theta \end{bmatrix} + \begin{bmatrix} V_{\tau x} \\ Z_{\theta x} \end{bmatrix} \delta x \right] \tag{7.83}$$

Note that this requires the inversion of a matrix with determinant D given by

$$D = Z_{\tau\tau} Z_{\theta\theta} - Z_{\theta\tau} Z_{\tau\theta} \tag{7.84}$$

If terminal constraints are present it may be possible to eliminate one or more of the associated Lagrange multipliers with the change in switching time $d\tau$ and the change in the parameters $\delta\theta$.

7.6.4 Jump Conditions for the Partial Derivatives

As might be expected, discontinuities are introduced into the partial derivatives of the return function. The relationships that are obtained are very similar to those given by Equation (7.34) and will not be derived here except in the special case when the nominal trajectory is the optimal trajectory. In this case $Z_\tau = 0$ and $Z_\theta = 0$ and Equation (7.83) becomes

$$\begin{bmatrix} d\tau \\ \delta\theta \end{bmatrix} = -\begin{bmatrix} Z_{\tau\tau} & Z_{\tau\theta} \\ Z_{\theta\tau} & Z_{\theta\theta} \end{bmatrix} \begin{bmatrix} V_{\tau x} \\ Z_{\theta x} \end{bmatrix} \delta x \tag{7.85}$$

The substitution of this equation back into Equation (7.80) yields the following relations between partial derivatives of the return function on either side of the discontinuity:

$$\begin{aligned} V_x^- &= V_x^+ \\ V_{xx}^- &= V_{xx}^+ - [V_{x\tau} \quad Z_{x\theta}] \begin{bmatrix} Z_{\tau\tau} & Z_{\tau\theta} \\ Z_{\theta\tau} & Z_{\theta\theta} \end{bmatrix}^{-1} \begin{bmatrix} V_{\tau x} \\ Z_{\theta x} \end{bmatrix} \end{aligned} \tag{7.86}$$

7.7 TESTS FOR OPTIMALITY

If an impulse occurs at an intermediate time τ, then the stationarity conditions that must be satisfied by an optimal impulse are

$$Z_\theta = V_x^+ h_\theta + P_\theta = 0 \tag{7.87}$$

$$V_\tau = V_x^+(f^- - f^+) + L^- - L^+ = 0 \tag{7.88}$$

The convexity condition for a maximum is

$$\begin{bmatrix} Z_{\theta\theta} & Z_{\theta\tau} \\ Z_{\tau\theta} & Z_{\tau\tau} \end{bmatrix} < 0 \tag{7.89}$$

where

$$\begin{aligned} Z_{\theta\theta} &= h_\theta^{\mathrm{T}} V_{xx} h_\theta + V_x h_{\theta\theta} + P_{\theta\theta} \\ Z_{\theta\tau} &= [(L_x^- - L_x^+) + V_{xx}^+(f^- - f^+) + V_x^+(f_x^- - f_x^+)]\, h_\theta + P_{\theta\theta} \\ Z_{tt} &= (f^- - f^+)^{\mathrm{T}} V_{xx}(f^- - f^+) - (V_x^+ f_x^+ + L_x^+)(f^- - f^+) \\ &\quad + (V_x^+(f_x^- - f_x^+) + L_x^- - L_x^+)\, f^- \\ &\quad + V_t^+(f_t^- - f_t^+) + L_t^- - L_t^+ + P_{tt} \end{aligned}$$

If the time of the impulse is fixed, then the conditions that must apply are

$$Z_\theta = 0 \tag{7.90}$$

$$Z_{\theta\theta} < 0 \tag{7.91}$$

It is also necessary to check whether or not the addition of an impulse will improve the performance index. To do this, we must check $Z_\theta \, d\theta < 0$ for admissible $d\theta$ at some intermediate time.

7.8 EXAMPLE

A particularly important application of this analysis is the computation of midcourse correction maneuvers in interplanetary space missions. The correction is affected by an impulsive change in the velocity of the spacecraft.

For this example, consider a spacecraft moving in a gravitational field and constrained to move in a plane. The position (x_1, x_2) and velocity (x_3, x_4) are given by cartesian coordinates as in Example 7.3. The impulsive corrections change the velocity (x_3, x_4) as follows,

$$x_3^+ = x_3^- + \theta_1 \cos \theta_2$$

$$x_4^+ = x_4^- + \theta_1 \sin \theta_2$$

where θ_1 is the magnitude of the correction and, θ_2 is the angle the impulse makes with the x_1 axis. Thus, $\theta_1 \geqslant 0$ and $0 \leqslant \theta_2 \leqslant 2\pi$. The problem is to choose the correction so that a desired terminal state is reached while minimizing the amount of fuel (thrust, θ_1) used. A suitable performance index to be minimized is

$$J = (x - x_f) \, \Lambda(x - x_f) + \sigma\theta_1$$

The first term ensures that certain terminal constraints are met to the desired accuracy while the second term weights the thrust. The matrix Λ and the weighting parameter σ adjust the relative importance of the terms in the index. For this example, the elements of Λ were chosen as $\Lambda_{ij} = 0$ except $\Lambda_{11} = 1$, $\Lambda_{22} = 1$, and $\sigma = 0.01$ and the terminal time was free. The initial state was chosen as $x_1 = 1.0$, $x_2 = 0.0$, $x_3 = 0.0$, and $x_4 = 1.0$, and the final state x_f as $x_1(t_f) = 0.0$, $x_2(t_f) = 1.2$. The final velocities $x_3(t_f)$ and $x_4(t_f)$ were not weighted in the performance index.

The partial derivatives needed to implement the sweep algorithm were then formed. The procedure may be summarized as follows:

1. Choose priming values for the control variables. Here the values $\theta_1 = 0.12$, $\theta_2 = 1.7$ rad, $t_f = 1.6$, and $\tau_1 = 0.25$ were chosen.

2. Integrate the state equations forward and store the state at the switching time τ_1 and the final time t_f.

3. Compute the change in the final time as described in Chapter 6.

4. Integrate the equations for V_x and V_{xx} backward from the final time until the switching time.

5. Compute and store the quantities Z_τ, Z_θ, etc., Equation (7.79).

6. (Note that in this type of problem it is not necessary to integrate the equations for V_x, etc., past the initial discontinuity.) Integrate the state equations forward to the first discontinuity. Compute $d\tau$ and $\delta\theta$ from Equation (7.83) and from $\tau_1^{j+1} = \tau_1^j + d\tau$ and $\theta^{j+1} = \theta^j + \delta\theta$.

7. Continue the integration to the final time.

8. Repeat steps 3–8 until no further improvement can be obtained.

In this problem convergence was obtained in 7 iterations (see Table XXX). Other quantitites associated with the convergence are also shown in the table. As might be anticipated from the low thrust example in Chapter 2, the optimal time for the impulse was the initial time. The tests for optimality of the impulse were also applied and shown to hold. Since the impulse was at the initial time, it was necessary to verify that $Z_\theta = 0$, $Z_\tau > 0$, and $Z_{\theta\theta} > 0$. This condition did hold for the final solution.

Checks for intermediate impulses were also made. Z_θ had the form

$$Z_{\theta_1} = \sigma + V_{x_3} \cos \theta_2 + V_{x_4} \sin \theta_2$$

$$Z_{\theta_2} = -(V_{x_3} \sin \theta_2 + V_{x_4} \cos \theta_2)\, \theta_1$$

Now a nontrivial impulse will occur if and only if θ_1 should be different from zero, i.e., $Z_{\theta_1}\, d\theta_1 <$ or $Z_{\theta_1} < 0$. Now Z_{θ_1} is only a function θ_2, thus, choosing θ_2 to minimize Z_{θ_1} yields $Z_{\theta_1} = \sigma - V_{x_3}^2 + V_{x_4}^2$. If Z_{θ_1} is negative, then an additional impulse should be added. Thus a necessary condition is that $0 < \sigma - V_{x_3}^2 + V_{x_4}^2$ hold along the optimal solution. This condition held along the final solution.

PROBLEMS

1. For the continuous state variable problem with terminal constraints, derive a local stationarity condition for the switching time.
2. Repeat the analysis of Section 7.6.1 for the general discontinuity $x^+ = g(x^-, \theta, \tau)$.
3. Derive the successive sweep equations for an impulse at the final time when the final time is fixed. What are the necessary conditions that must hold?
4. Derive the successive sweep equation for an impulse at the final time when the final time is free. What are the necessary conditions that must hold?

BIBLIOGRAPHY AND COMMENTS

The material in this chapter was first presented by the authors in

Dyer, P. and McReynolds, S. R. (1968). "Optimal Control Problems with Discontinuities," *J. Math. Anal. Appl.*, Vol. 23, No. 3, p. 585.

Dyer, P. and McReynolds, S. R. (1969). "Optimization of Control Problems with Discontinuities and Terminal Constraints," *IEEE Trans. Autom. Control*, Vol. AC-14, No. 3, p. 223.

Similar results have been obtained by Jacobson:

Jacobson, D. H. (1968). "Differential Dynamic Programming Algorithm for Solving Bang-Bang Control Problems," *IEEE Trans. Autom. Control*, Vol. AC-13, No. 6, pp. 661–675.

A large volume of literature exists in the area of bang-bang control and impulsive guidance. For a list of references, see

Athans, M. (1966). "The Status of Optimal Control Theory and Applications for Deterministic Systems," *IEEE Trans. Autom. Control*, Vol. AC-11, No. 3, pp. 580–599.

8

The Maximum Principle and the Solution of Two-Point Boundary Value Problems

8.1 INTRODUCTION

In this chapter the continuous optimal control problem will be reformulated in terms of a two-point boundary value problem. A two-point boundary value problem is, apparently, a totally different mathematical problem than the initial value problem in partial differential equations developed via dynamic programming in the previous chapters. Computationally this difference becomes apparent in differing approaches to the development of numerical algorithms.

The two-point boundary value problem arises in the classical calculus of variations approach to optimal control problems via an application of the maximum principle.

The maximum principle consists of a set of necessary (although in general *not* sufficient) conditions that an optimal solution must satisfy, for example, the Euler–Lagrange equations, the transversality conditions, Weirstrasse's E-condition, etc. Although in theory multiple extrema can exist, in practice such complications do not arise too often; when they do, however, the foregoing conditions may be used to determine a unique optimum.

Traditionally the maximum principle has been derived using classical variational theory. However, here the maximum principle is derived from the dynamic programming equations with the help of the theory of characteristics. It will be seen that the two-point boundary value problem is essentially equivalent to the set of stationarity conditions derived in Chapter 5. For the linear quadratic problem these equations are shown to constitute a linear two-point boundary problem.

Next, several methods are presented for solving such linear two-point boundary value problems: the superposition method, the adjoint solution technique, and the sweep method. The sweep method is shown to be equivalent to the algorithm developed in Chapter 4 directly from the dynamic programming equations. The stability of these methods is examined and the superiority of the sweep method is demonstrated.

Further sections discuss the application of Newton–Raphson techniques for solving nonlinear problems. It will be seen that the different algorithms can be classified into two different categories: shooting methods, which iterate on boundary conditions, and quasi-linearization methods, which iterate on functions.

Finally, the subject of invariant imbedding is discussed and used to derive the successive sweep algorithm. Conceptually it will be seen that invariant imbedding is similar to dynamic programming. A general set of conditions is derived under which the invariant imbedding equations may be converted into a single dynamic programming equation.

8.2 THE MAXIMUM PRINCIPLE

The maximum principle will be developed by applying the method of characteristics to the dynamic programming equations. To simplify the analysis, only the basic problem (without control parameters or side constraints) will be considered. The problem is to find the control functions $u(t)$, $t_0 \leqslant t \leqslant t_f$, and maximize the performance index J

$$J = \phi(x(t_f), t_f) + \int_{t_0}^{t_f} L(x, u, t)\, dt \tag{8.1}$$

where the state x is given by

$$\dot{x} = f(x, u, t) \tag{8.2}$$

$$x(t_0) = x_0 \tag{8.3}$$

As was shown in Chapter 3 the dynamic programming solution consists of finding the functions $V^{\text{opt}}(x, t)$ and $u^{\text{opt}}(x, t)$ that satisfy the following boundary value problem:

$$u^{\text{opt}}(x, t) = \arg\max_u \{V_x^{\text{opt}}(x, t)\, f(x, u, t) + L(x, u, t)\} \tag{8.4}$$

$$V^{\text{opt}}(x, t_f) = \phi(x, t_f) \tag{8.5}$$

$$0 = V_t^{\text{opt}}(x, t) + V_x^{\text{opt}}(x, t)\, f(x, u^{\text{opt}}, t) + L(x, u^{\text{opt}}, t) \tag{8.6}$$

The optimal solution $u^{\text{opt}}(x, t)$ corresponding to the given initial condition is then obtained in a feedback process by solving the initial value problem

$$\dot{x}^{\text{opt}}(t) = f(x^{\text{opt}}(t), u^{\text{opt}}(t), t)$$
$$x^{\text{opt}}(t_0) = x_0 \tag{8.7}$$

$$u^{\text{opt}}(t) = u^{\text{opt}}(x^{\text{opt}}(t), t) \tag{8.8}$$

where $x^{\text{opt}}(t)$ denotes the optimal trajectory corresponding to the optimal control and the specified initial conditions x_0.

Note that Equation (8.4) implies that the optimal control $u^{\text{opt}}(x, t)$ may be determined as a function of the state x, the time t, and the partial derivative of the return V_x^{opt}. For example, $u^{\text{opt}}(x, t)$ might be found by solving the stationarity equation, which is obtained by differentiating the right-hand side of Equation (8.4) and setting this derivative to zero, i.e., u^{opt} is chosen such that

$$V_x^{\text{opt}} f_u + L_u = 0 \tag{8.9}$$

Hence, in order to obtain the optimal control for a particular set of initial conditions it is necessary only to determine V_x^{opt} along this trajectory. A characteristic equation for V_x^{opt} is obtained by differentiating Equations (8.5) and (8.6) with respect to x. Differentiating Equation (8.5) gives

$$V_x^{\text{opt}}(x, t_f) = \phi_x(x, t_f) \tag{8.10}$$

Thus V_x^{opt} is obtained at the final time as a function of $x(t_f)$ and t_f. Differentiating Equation (8.6) yields

$$\begin{aligned} 0 = {} & V_{tx}^{\text{opt}}(x, t) + V_{xx}^{\text{opt}}(x, t)\, f(x, u^{\text{opt}}(x, t), t) \\ & + V_x^{\text{opt}}(x, t)\, f_x + L_x \\ & + [V_x^{\text{opt}}(x, t)\, f_u + L_u](\partial u^{\text{opt}}/\partial x) \end{aligned} \tag{8.11}$$

This expression, Equation (8.11), may be simplified by noting that Equation (8.9) implies that the last term is identically zero. Also it may be noted that

$$V_{tx}^{\text{opt}}(x, t) + V_{xx}^{\text{opt}}(x, t)\, f(x, u^{\text{opt}}(x, t), t) = \dot{V}^{\text{opt}}$$

represents the *total* time derivative of $V_x(x, t)$ along a trajectory determined by $u^{\text{opt}}(x, t)$. Thus Equation (8.11) may be written

$$\dot{V}_x^{\text{opt}} = -V_x^{\text{opt}} f_x(x^{\text{opt}}, u^{\text{opt}}(t), t) - L_x(x^{\text{opt}}, u^{\text{opt}}(t), t) \tag{8.12}$$

The derivation of the relations needed to form the maximum principle is now complete. However, in order to summarize the principle in the conventional notation the following notation will be employed

$$p(t) = V_x^{\text{opt}}(x^{\text{opt}}(x, t)) \tag{8.13}$$

$$H(x, u, p, t) = p(t)\, f(x, u, t) + L(x, u, t) \tag{8.14}$$

The n-vector p is referred to as the *adjoint* or *costate* vector, or sometimes as the vector of Lagrange multipliers. The function H is referred to as the *Hamiltonian.*

With this notation the maximum principle can be summarized as follows. In order that $u^{\text{opt}}(t)$ be the optimal solution for this problem, it is necessary that (1) there exists a function $p^*(t)$ corresponding to $x^{\text{opt}}(t)$ and $u^{\text{opt}}(t)$ such that $p^*(t)$ and $x^{\text{opt}}(t)$ satisfy the equations

$$\begin{aligned} \dot{x}^{\text{opt}} &= f(x^{\text{opt}}, u^{\text{opt}}(t), t) = H_p(x^{\text{opt}}, u^{\text{opt}}, p^*, t) \\ \dot{p}^* &= -H_x(x^{\text{opt}}, u^{\text{opt}}, p^*, t) \end{aligned} \tag{8.15}$$

with the boundary conditions

$$x^{\text{opt}}(t_0) = x_0 \tag{8.16a}$$

$$p^*(t_f) = \phi_x(x^{\text{opt}}, t_f) \tag{8.16b}$$

and (2) the function $H(x^{\text{opt}}, p^*, u, t)$ has an absolute maximum as a function u at $u = u^{\text{opt}}$ for all t in $[t_0, t_f]$; i.e.,

$$\max_u H(x^{\text{opt}}, p^*, u, t) = H(x^{\text{opt}}, p^*, u^{\text{opt}}, t) \tag{8.17}$$

This last condition may often be used to determine u^{opt} and the condition

$$H_u(x^{\text{opt}}, p^*, u^{\text{opt}}, t) = 0 \tag{8.18}$$

will often hold. Equations (8.15), (8.16), and (8.18) are exactly the stationarity conditions that were developed in Chapter 5 for a local maximum.

The extension of the maximum principle to problems with control parameters is straightforward and will not be given here. Side constraints may be included by using the Lagrange multiplier technique. For example, if terminal constraints are present, i.e.,

$$\psi(x, t_f) = 0 \tag{8.19}$$

the constraints are adjoined to the performance index equation (8.1) with multipliers ν. The equation for $p(t_f)$ must then be modified to

$$p(t_f) = \phi_x(x, t_f) + \nu^{\mathrm{T}}\psi_x(x, t_f)$$

where the multipliers ν may treated as unknown parameters.

Note that in any case Equations (8.15), (8.16), or (8.18) constitute a two-point boundary value problem. In general the solution of this problem is difficult, although as might be expected a straightforward solution to the linear quadratic problem does exist.

8.3 THE LINEAR QUADRATIC PROBLEM

Following the notation of Chapter 4 the quadratic performance index is given by

$$J = \tfrac{1}{2}x(t_f)^{\mathrm{T}}\,\phi_{xx}x(t_f) + \tfrac{1}{2}\int_{t_0}^{t_f} \{x^{\mathrm{T}}L_{xx}x + u^{\mathrm{T}}L_{uu}u\}\,dt \tag{8.20}$$

and the state equations are

$$\dot{x} = f_x x + f_u u, \qquad x(t_0) = x_0 \quad \text{specified} \tag{8.21}$$

Thus the Hamiltonian H becomes

$$H = \tfrac{1}{2}x^{\mathrm{T}}L_{xx}x + \tfrac{1}{2}u^{\mathrm{T}}L_{uu}u + p^{\mathrm{T}}(f_u x + f_u u) \tag{8.22}$$

and so u^{opt} must be chosen such that [cf Equation (8.18)]

$$u^{\mathrm{T}}L_{uu} + p^{\mathrm{T}}f_u = 0,$$

i.e.,

$$u^{\mathrm{opt}}(p, x) = -L_{uu}^{-1}f_u^{\mathrm{T}}\,p \tag{8.23}$$

This equation is now used to eliminate u from the Hamiltonian in Equation (8.22), giving

$$H^{\mathrm{opt}} = \tfrac{1}{2}x^{\mathrm{T}}L_{xx}x + p^{\mathrm{T}}f_x x - \tfrac{1}{2}p^{\mathrm{T}}f_u L_{uu}f_u^{\mathrm{T}}p \tag{8.24}$$

Hence the two-point boundary value problem, Equations (8.15) and (8.16), is defined by

$$\dot{x} = f_x x - f_u L_{uu}^{-1}f_u^{\mathrm{T}}\,p, \qquad \dot{p} = -L_{xx}x - f_x^{\mathrm{T}}\,p \tag{8.25}$$

and

$$x(t_0) = x_0\,, \qquad p(t_f) = \phi_{xx}x(t_f) \tag{8.26}$$

The interesting feature of this two-point boundary value problem is that both the differential equations [Equation (8.25)] and the boundary conditions [Equation (8.26)] *are linear*. Such linear problems have received a great deal of attention and several straightforward methods exist for their solutions. In the next section three of these methods will be examined.

8.4 TECHNIQUES FOR SOLVING LINEAR TWO-POINT BOUNDARY VALUE PROBLEMS

In this section three techniques for solving linear two-point boundary value problems will be discussed, namely, (1) superposition, (2) adjoint solutions, and (3) the sweep method. For the purposes of the discussion, the two-point boundary value problem derived in the previous section will be considered in a more generalized form, i.e.,

$$\dot{x} = Ax + Bp + v, \qquad \dot{p} = -Cx - A^{\mathrm{T}}p - w \tag{8.27}$$

$$x(0) = x_0 \tag{8.28}$$

$$p(t_f) = \phi_{xx}(t_f)\, x(t_f) \tag{8.29}$$

where [cf Equations (8.25)]

$$A = f_x, \qquad B = -f_u L_{uu}^{-1} f_u^{\mathrm{T}}, \qquad C = L_{xx} \tag{8.30}$$

8.4.1 Superposition

A fundamental $2n \times n$ matrix solution $Y(t)$ is obtained for the differential equations (8.27) with $v = 0$ and $w = 0$. Also a single solution $y(t)$ is obtained with the inhomogeneous equations (8.27). Thus $Y(t)$ is given by

$$\dot{Y} = \begin{bmatrix} A & B \\ -C & -A^{\mathrm{T}} \end{bmatrix} Y(t) \tag{8.31}$$

$$Y(t_0) = \underbrace{\begin{bmatrix} 0 \\ \cdots \\ I \end{bmatrix}}_{n} \begin{matrix} \updownarrow n \\ \updownarrow n \end{matrix} \tag{8.32}$$

and $y(t)$ is obtained from

$$\dot{y} = \begin{bmatrix} A & B \\ -C & -A^{\mathrm{T}} \end{bmatrix} y(t) + \begin{bmatrix} v \\ -w \end{bmatrix} \tag{8.33}$$

$$y(t_0) = \begin{bmatrix} x_0 \\ \cdots \\ 0 \end{bmatrix} \begin{matrix} n \\ n \end{matrix} \tag{8.34}$$

A general solution $x(t)$ and $p(t)$ to Equation (8.27) that satisfies the initial boundary condition (8.28) can be written as

$$\begin{bmatrix} x(t) \\ p(t) \end{bmatrix} = Y(t)\, p_0 + y(t) \tag{8.35}$$

where p_0 is a set of n-constants that correspond to a certain choice of initial conditions for $p(t)$. They are determined at the final time from the terminal boundary condition equation (8.29)

$$0 = \begin{bmatrix} I \\ -\phi_{xx}(x_f, t_f) \end{bmatrix} [Y(t_f)\, p_0 + y(t_f)] \tag{8.36}$$

The solution of this equation (8.36) determines the missing initial condition on p_0. A solution to the linear two-point boundary value problem may be obtained by reintegrating the differential equations forward with the initial conditions x_0, p_0.

8.4.2 The Method of Adjoint Equations

The method of adjoint equations is based upon generating a fundamental set of adjoint solutions, which will be denoted by $Y^*(t)$ and $y^*(t)$. The solution $Y^*(t)$, an $n \times 2n$ matrix, is given by

$$Y^{*\mathrm{T}} = -\begin{bmatrix} A^{\mathrm{T}} & -C^{\mathrm{T}} \\ B^{\mathrm{T}} & -A \end{bmatrix} Y^{*\mathrm{T}}, \qquad Y^{*\mathrm{T}}(t_f) = \begin{bmatrix} \phi_{xx} \\ -I \end{bmatrix} \tag{8.37}$$

and the solution $y^*(t)$, an n-vector, is given by

$$\dot{y}^* = -Y^*(t) \begin{bmatrix} v \\ -w \end{bmatrix}; \qquad y^*(t_f) = 0$$

A general solution $x(t)$ and $p(t)$ to Equation (8.27) that satisfies the terminal boundary condition equation (8.29) satisfies the equation

$$0 = Y^*(t) \begin{bmatrix} x(t) \\ p(t) \end{bmatrix} + y^*(t) \tag{8.38}$$

The equations for $Y^*(t)$ and $y^*(t)$ are usually constructed by seeking functions such that the relation (8.38) agrees with (8.29) at the terminal time and such that the condition,

$$0 = \frac{d}{dt} \left\{ Y^*(t) \begin{bmatrix} x(t) \\ p(t) \end{bmatrix} + y^*(t) \right\} \tag{8.39}$$

holds. Thus the terminal boundary condition (8.29) may be satisfied by choosing $x(t)$ and $p(t)$ to satisfy (8.38) at any arbitrary time. In particular, a solution that satisfies the terminal boundary conditions and the initial boundary conditions may be determined at the initial time from

$$0 = Y^*(t_0)\begin{bmatrix} x_0 \\ p(t_0) \end{bmatrix} + y^*(t_0) \tag{8.40}$$

8.4.3 The Sweep Method

The sweep method also generates a fundamental solution to the linear two-point boundary value problem. It does so by generating an $n \times n$ matrix $P(t)$ and a vector $h(t)$ such that a general solution $x(t)$ and $p(t)$ to the differential equation (8.27), which satisfies the terminal boundary conditions (8.29), also satisfies the equation,

$$p(t) = P(t)\, x(t) + h(t) \tag{8.41}$$

Differential equations for $P(t)$ and $h(t)$ are chosen so that

$$(d/dt)(P(t)\, x(t) + h(t) - p(t)) = 0 \tag{8.42}$$

holds identically for all solutions to Equation (8.27). Carrying out the indicated differentiation on each term in Equation (8.42) gives

$$\dot{P}x + P\dot{x} + \dot{h} - \dot{p} = 0 \tag{8.43}$$

Using Equation (8.41) to eliminate p from Equation (2.29) and Equation (8.27) results in

$$\dot{x} = (A + BP)x + Bh + v \tag{8.44}$$

$$\dot{p} = -(C + A^{\mathrm{T}}P)x - A^{\mathrm{T}}h - w \tag{8.45}$$

Using these equations to eliminate $\dot{x}$ and $\dot{p}$ from (8.43) yields

$$(\dot{P} + PA + A^{\mathrm{T}}P + PBP + C)x + (\dot{h} + (PB + A^{\mathrm{T}})h + Pv + w) = 0 \tag{8.46}$$

This condition will hold for all x if $P(t)$ and $h(t)$ are chosen to satisfy

$$\dot{P} + PA + A^{\mathrm{T}}P + PBP + C = 0 \tag{8.47}$$

$$\dot{h} + (PB + A^{\mathrm{T}})h + Pv + w = 0 \tag{8.48}$$

Boundary conditions for $P(t)$ and $h(t)$ are supplied by equating Equation (8.41) to Equation (8.29) at the terminal time, i.e.,

$$P(t_f) = \phi_{xx}(x, t_f) \tag{8.49}$$

$$h(t_f) = 0 \tag{8.50}$$

In order to obtain the particular solution that satisfies both the initial and terminal boundary conditions, we may choose $p(t_0) = P(t_0)\, x_0 + h(t_0)$ and with x_0 integrate Equation (8.27) forward. Alternatively, it would be possible to integrate Equation (8.44) forward and use Equation (8.41) to obtain $p(t)$. This second technique has the advantage of being more stable as well as integrating fewer (n) equations. However, a large amount of computer storage is required mainly to store the $n \times n$ matrix $P(t)$ at each integration step.

Another approach is based on expressing u^{opt} as a function of that state $x(t)$, i.e.,

$$u^{\text{opt}}(t) = -L_{uu}^{-1}(f_u{}^{\mathrm{T}} P(t)\, x(t) + h(t)) \tag{8.51}$$

This expression may be used as a feedback law to generate the desired solution. The advantage here is that only the $m \times n$ matrix $L_{uu}^{-1} f_u{}^{\mathrm{T}} P(t)$ has to be stored. If the control variable has many fewer dimensions than the state, the saving can be significant.

8.4.4 Comparison of the Three Methods

The superposition method and the method of adjoint solutions are similar in that both require the integration of $2n^2$ solutions to the set of homogeneous linear differential equations obtained by setting $v = 0$, $w = 0$ in Equation (8.27). On the other hand, the sweep method requires the integration of a Riccati matrix equation (8.47). Since the equation is symmetric, the number of differential equations is only $n(n + 1)/2$. A solution to a set of n linear differential equations must also be integrated in all three cases.

Conceptually, the sweep method is more closely related to the method of adjoint solutions. Both generate a *field of equivalent boundary conditions* (Gelfand and Fomin, 1963). Note that the sweep method continually solves $p(t)$ as a function of $x(t)$, whereas the method of adjoint solutions performs this inversion only at the initial time. The reader should also note that the sweep method is exactly equivalent to the dynamic programming algorithm for the linear quadratic problem given in Chapter 3. Both algorithms require the integration of the same matrix Riccati equation.

The methods of adjoint solutions and superposition may run into difficulty if the matrix that requires inversion is singular (or near singular). If either situation occurs, a *conjugate point* (see Appendix A) has been encountered and the second variation has a zero eigenvalue. If a conjugate point exists anywhere on the interval $[t_0, t_f]$, the sweep method will detect it by the fact that the solution to the Riccati equation blows up. Thus, the sweep method has the advantage that it detects intermediate conjugate points, whereas either superposition or the method of adjoint solutions detect only conjugate points at the ends of the trajectory.

8.4.5 Stability of the Numerical Methods

An important property of any numerical method that must be considered is its stability. A numerical algorithm will be said to be stable if numerical errors due to round-off and truncation do not increase. In comparing the numerical algorithms used to solve linear two-point boundary value problems the growth of these errors over the length of the trajectory must be examined. In each algorithm the basic numerical computation involves the integration of a set of differential equations. Thus the numerical stability of the algorithm depends upon the stability of these differential equations.

Before proceeding to analyze the stability of these algorithms the idea of stability will be defined more precisely.

DEFINITION 8.1. Consider the set of differential equations $\dot{y}(t) = Dy(t)$ where D is an $m \times m$ constant matrix. The system is said to be strictly stable if, for every solution $y(t)$ of the system

$$\| y(t) \| \to 0 \qquad \text{as} \quad t \to \infty$$

Hence, the system is strictly stable provided the eigenvalues of the matrix D are negative.†

Next it is shown that the methods of superposition and adjoint solution are not strictly stable in the sense just defined. For both methods the matrix D has the form [cf Equation (8.37)],

$$D = \begin{bmatrix} A & B \\ -C & -A^{\mathrm{T}} \end{bmatrix}$$

† Similar results could be obtained for time-varying systems, although the analysis is somewhat complicated. However, generally speaking the eigenvalues are required to be negative to ensure strict stability.

where B and C are symmetric and A, B, and C are $n \times n$ matrices. Now note that $SDS^{\mathrm{T}} = -D^{\mathrm{T}}$ if S is chosen to be

$$S = \begin{bmatrix} 0 & I \\ -I & 0 \end{bmatrix}$$

As S is an orthonormal matrix the eigenvalues of D are unchanged by the foregoing rotation and hence are equal to the eigenvalues of $-D^{\mathrm{T}}$. But the eigenvalues of a square matrix are unchanged if the matrix is transposed. Thus the eigenvalues of D must occur in pairs with opposite signs.† Hence not all of the real parts of the eigenvalues of D can be negative and the system is not strictly stable.

The stability properties of the backward sweep may be examined by considering a perturbation of Equations (8.47) and (8.48) in δP and δh, i.e.,

$$\delta\dot{P} + \delta P(A + BP) + (A^{\mathrm{T}} + PB)\,\delta P = 0 \tag{8.52}$$

$$\delta\dot{h} + (A^{\mathrm{T}} + PB)\,\delta h + \delta P v + w = 0 \tag{8.53}$$

The previous equations will be stable or unstable depending on the eigenvalues of the matrix $[A^{\mathrm{T}} + PB]$. Both equations are stable if the real parts of the eigenvalues of $(A^{\mathrm{T}} + PB)$ are negative.

Now, if the corrective feedback law is used, Equation (8.44) is integrated forward. Note that the stability here depends on the matrix $(A + BP) = (A^{\mathrm{T}} + PB^{\mathrm{T}})^{\mathrm{T}}$; thus the forward equations of the sweep method are stable if the equations on the backward part of the sweep are stable. So if the field of optimal solutions is stable in the sense that a small change in the initial state is not magnified, then the successive sweep method is stable in a region containing the optimum. In most practical cases the desired optimal solution will be strictly stable and so the sweep method will also be stable.‡ Because of its stability properties the sweep method has an inherent advantage over the superposition and adjoint methods.

These results will now be illustrated by an example. Consider the problem of maximizing the following performance index

$$J = -\tfrac{1}{2}x^2(t_f) - \tfrac{1}{2}\int_{t_0}^{t_f} (a^2x^2 + u^2)\,dt \tag{8.54}$$

† This proof is the work of Broucke and Lass (1968).

‡ Some general conditions have been established under which the Riccati equation is asymptotically stable [cf Kalman (1960) and Lee and Markus (1967), Chapter 4].

where the system equations are given by

$$\dot{x} = f_x x + f_u u \tag{8.55}$$

and where x and u are scalars. The nominal solution $u(t)$ and $x(t)$ will be chosen to be identically zero. The Hamiltonian H is given by

$$H = p^{\mathrm{T}} f_x + p^{\mathrm{T}} f_u - \tfrac{1}{2}(a^2 x^2 + u^2) \tag{8.56}$$

where p is the adjoint variable.

The nominal p^0 satisfies the equation

$$p^0(t_1) = 0, \qquad \dot{p}^0(t) = -f_x p^0 + a^2 x \tag{8.57}$$

and hence

$$p^0(t) = 0, \qquad t \in [t_1, t_f] \tag{8.58}$$

The quantities A, B, C, v, and w of Equation (8.27) are given by

$$A = f_x, \qquad B = f_u^2, \qquad C = -a^2 \tag{8.59}$$

$$v = 0, \qquad w = 0 \tag{8.60}$$

The backward equations for the sweep method are

$$P(t_1) = 1; \qquad \dot{P} = -2Pf_x - P^2 f_u^2 + a^2 \tag{8.61}$$

$$h(t_1) = 0; \qquad \dot{h} = -(Pf_x + f_u^2)h \tag{8.62}$$

and hence the solution for h is identically zero. The backward equations for the method of adjoint solutions are

$$Y^{*\mathrm{T}}(t_1) = \begin{bmatrix} -I \\ I \end{bmatrix} \dot{Y}^{*\mathrm{T}} = -\begin{bmatrix} f_x & a^2 \\ f_u^2 & -f_x \end{bmatrix} Y^{*\mathrm{T}} \tag{8.63}$$

The fact that $v = w = 0$ implies $y^*(t) = 0$. Now let us consider the stability of these two systems of differential equations. The eigenvalues of the equation for Y^* are given by

$$\mu_1, \mu_2 = \pm(f_u^2 a^2 + f_x^2)^{1/2} \tag{8.64}$$

and hence one eigenvalue is greater than zero and one is smaller than zero. Thus the method of adjoint solutions (and hence the method of superposition), is unstable for this example. Instability also effects both directions of integration for each iteration. Note that the degree of instability increases as f_x becomes large, regardless of the sign. This suggests why the method of adjoint solutions, or superposition, is unsatisfactory for highly damped systems (where $f_x \ll 0$).

Now let us examine the stability of the sweep method more closely. The perturbation of the equation on the backward sweep is

$$\delta \dot{P} = -2(f_x^{\mathrm{T}} + P f_u^2)\, \delta P \tag{8.65}$$

which is stable provided the matrix $[f_x^{\mathrm{T}} + P f_u]$ has negative eigenvalues. Figure 8.1 displays a graph of $-\dot{P}$ versus P. The regions of stability and

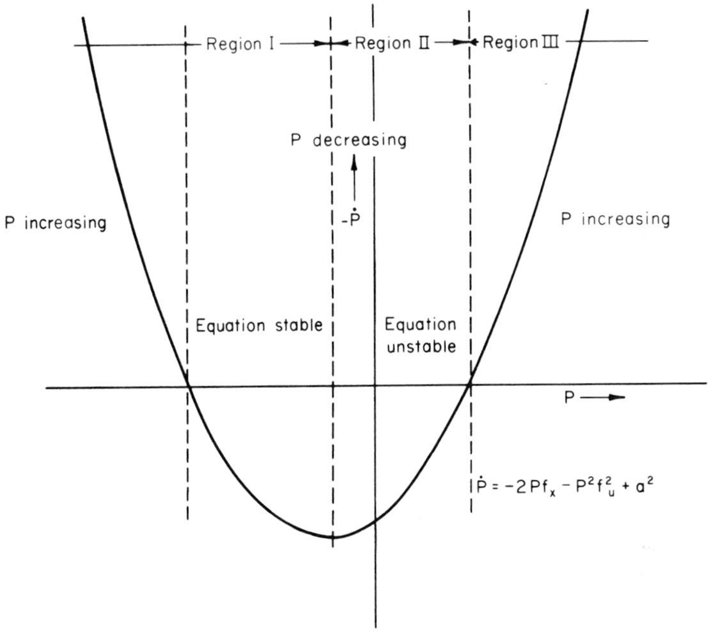

FIG. 8.1. The Riccati equation, $\dot{P} = -2Pf_x - P^2 f_u^2 + a^2$.

instability are indicated. Arrows are also used to indicate where the solution is increasing or decreasing on the backward integrations. From the graph we conclude that there are three regions:

Region I. The Riccati equation is stable and $P(t)$ approaches an asymptotic value.

Region II. The Riccati equation is *locally unstable*, but $P(t)$ will eventually become stable and approach an asymptotic value.

Region III. The Riccati equation is unstable and $P(t)$ will approach $+\infty$.

For this example, since $P(t_f) = -1$, the Riccati equation will become stable and will also have the property of asymptotic stability.

Note that if $f_u = 0$, the Riccati equation degenerates into a linear equation that is stable if the system equations are stable.

8.5 NEWTON–RAPHSON METHODS FOR SOLVING NONLINEAR TWO-POINT BOUNDARY VALUE PROBLEMS

Unfortunately elegant solutions are not possible for general nonlinear two-point boundary value problems. However, many iterative algorithms have been developed. In this section several of the second-order algorithms, based on the Newton–Raphson concept, will be discussed. These algorithms may be derived directly by considering a second-order expansion of the performance index. Here the indirect approach, based on a linearization of the two-point boundary value problem defined by the maximum principle, is preferred.

The unknowns are the functions $x(t)$, $p(t)$, $u(t)$, $t \in (t_0, t_f)$. The Newton–Raphson approach is based on generating a sequence of nominal solutions $\{x^j(t), p^j(t), u^j(t)\}$ which converges to the optimal solution $\{x^{\text{opt}}(t), p^{\text{opt}}(t), u^{\text{opt}}(t)\}$. Corrections to the current nominal are obtained by linearizing the nonlinear boundary value problem about the nominal solution and solving the resultant linear problem.

Now by introducing the vector functions $g(x, u, t)$, $h(p, x, u, t)$, $k(p, x, u, t)$, $m(p(t_f), x(t_f))$, and $n(x(t_0))$, the maximum principle can be expressed as

$$g(x, u, t) \equiv f(x, u, t) - \dot{x} = 0 \tag{8.66}$$

$$h(p, x, u, t) \equiv -H_x - \dot{p} = 0 \tag{8.67}$$

$$k(p, x, u, t) \equiv H_u = 0 \tag{8.68}$$

$$m(p(t_f), x(t_f)) \equiv \phi_x(t_f) - p(t_f) = 0 \tag{8.69}$$

and

$$n(x(t_0)) \equiv x_0 - x(t_0) = 0 \tag{8.70}$$

At a *nominal, nonoptimal* solution (x^k, p^k, u^k), some or all of these equations are not satisfied. Applying the Newton–Raphson approach, corrections $(\delta x, \delta p, \delta u)$ are obtained by solving linearized approximations to the equations (8.66)–(8.70). That is to say, the corrections are obtained from

$$\begin{aligned}
g_x\,\delta x + g_u\,\delta u &= -\epsilon g(x^k, u^k, p^k, t)\\
h_x\,\delta x + h_u\,\delta u + h_p\,\delta p &= -\epsilon h(x^k, u^k, p^k, t)\\
k_x\,\delta x + k_u\,\delta u + k_p\,\delta p &= -\epsilon k(x^k, u^k, p^k, t)\\
m_x\,\delta x(t_f) + m_p\,\delta p(t_f) &= -\epsilon m(x(t_f), p(t_f))\\
n_x\,\delta x(t_0) &= -\epsilon n(x^k(t_0))
\end{aligned} \tag{8.71}$$

These equations are the same as

$$\begin{aligned} \delta\dot{x} &= f_x\,\delta x + f_u\,\delta u + \epsilon g \\ \delta\dot{p} &= -H_{xx}\,\delta x - H_{xu}\,\delta u - H_{xp}\,\delta p + \epsilon h \\ 0 &= H_{ux}\,\delta x + H_{uu}\,\delta u + H_{up}\,\delta p + \epsilon k \end{aligned} \tag{8.72}$$

$$\begin{aligned} \delta p(t_f) &= \phi_{xx}\,\delta x(t_f) + \epsilon m \\ \delta x(t_0) &= \epsilon n \end{aligned} \tag{8.73}^{\dagger}$$

In order to convert the foregoing equations into the linear two-point boundary value problem considered earlier in this chapter, the third equation is used to eliminate δu, i.e.,

$$\delta u = -H_{uu}^{-1}(H_{ux}\,\delta x(t) + H_{up}\,\delta p(t) + \epsilon k(t)) \tag{8.74}$$

Now eliminating δu from the other equations, we obtain the following linear two-point boundary value problem

$$\begin{aligned} \delta\dot{x} &= A\,\delta x + B\,\delta p + v \\ \delta x(t_0) &= \epsilon n \\ \delta\dot{p} &= -C\,\delta x - A^{\mathrm{T}}\,\delta p - w \\ \delta p(t_f) &= \phi_{xx}\,\delta x(t_f) + \epsilon m \end{aligned} \tag{8.75}$$

where

$$\begin{aligned} A &= f_x - f_u H_{uu}^{-1} H_{ux} \\ B &= -f_u H_{uu}^{-1} f_u \\ C &= H_{xx} - H_{xu} H_{uu}^{-1} H_{ux} \\ v &= -\epsilon f_u H_{uu}^{-1} k + \epsilon g \\ u &= \epsilon H_{xu} H_{uu}^{-1} k - \epsilon h \end{aligned} \tag{8.76}$$

This linear two-point boundary value problem, Equation (8.75), must now be solved to obtain the corrections $(\delta x, \delta u, \delta p)$. These corrections may then be added to the nominal to obtain the new $(k + 1\text{th})$ nominal, i.e.,

$$\begin{aligned} x^{k+1}(t) &= x^k(t) + \delta x(t) \\ u^{k+1}(t) &= u^k(t) + \delta u(t) \\ p^{k+1}(t) &= p^k(t) + \delta p(t) \end{aligned}$$

† The parameter ϵ appearing in Equations (8.72) and (8.75) is the usual step-size parameter used to ensure the validity of the solutions by limiting the size of the corrections

Specific Newton–Raphson algorithms are obtained by constructing nominal solutions that *explicitly* satisfy some of the constraints. In effect this reduces the number of independent variables. Thus the differing versions of the Newton–Raphson algorithm may be classified by the independent variables, which are determined by the iterative procedure. Some of the special versions of the algorithm are as follows:

1. $p(t_0)$ is considered the independent variable (a shooting method). The variables $x^k(t_0)$, $x^k(t)$, $p^k(t)$, and $u^k(t)$ are chosen to satisfy Equations (8.66)–(8.68), and (8.70) by integrating Equations (8.66) forward with an arbitrary initial condition $p(t_0)$. Thus Equation (8.69) is the only equation that is not satisfied, and corrections must be found for $p(t_0)$.

2. $x(t_f)$ is considered the independent variable (a neighboring extremal method). This algorithm is similar to the first except that the final conditions are chosen so that the boundary condition at the final time is satisfied. $x(t)$, $u(t)$, and $p(t)$ are obtained by backward integration. The only condition that is not satisfied is the initial boundary condition (8.70).

3. $u(t)$ is considered the independent variable (a quasilinearization method). The variables $x(t_0)$, $x(t)$, $p(t_f)$, and $p(t)$ are chosen to satisfy (8.66), (8.67), (8.69), and (8.70). This is achieved by integrating the state equations forward and the costate equations *backward* using Equation (8.69) to initialize the backward integration for $p(t)$. Equation (8.68) is the only condition not satisfied.

4. $x(t)$ and $p(t)$ are considered the independent variables (a quasilinearization method). $u(t)$ is chosen to maximize the Hamiltonian, and hence Equation (8.68) is satisfied. Initially $x(t)$ and $p(t)$ are usually chosen so that the boundary condition equations (8.69) and (8.70) are satisfied. The differential equations (8.66) and (8.67) are relaxed on a nominal solution.

5. $x(t)$ and $u(t)$ are considered unknowns (a quasilinearization method). This is similar to the foregoing except that p is chosen by Equation (8.67) and $u(t)$ is not chosen by the maximum principle; hence, Equation (8.67) is satisfied but Equation (8.68) is not.

Conceptually the Newton–Raphson method can be divided into two different basic types: the shooting method and the quasilinearization method. Shooting methods, 1 and 2, treat boundary conditions as the equations that must be satisfied and the terminal value of the functions as the independent variable. Quasilinearization methods, 3, 4, and 5, treat functions as the independent variables and differential equations or an infinite set of algebraic equations [for example, Equation (8.68)] as the equations that must be satisfied. Shooting methods have a finite number of unknowns, whereas the quasilinearization method has an infinite number of unknowns. Shooting methods are simpler to implement. However, they have the disadvantage that solutions tend to be very sensitive to boundary conditions and hence numerical errors tend to hinder convergence.

Quasilinearization methods can be combined with Ritz methods to reduce the number of unknowns. For example, it may be assumed that the optimal control function can be approximated by $u(t) = \sum_{i=1}^{N} a_i S_i(t)$ where $[S_1(t), \ldots, S_N(t)]$ is a given set of functions, for example, $[1, t, t^2, t^3, \ldots, t^N]$.

Now the optimal choice of the parameters $[a_1, \ldots, a_n]$ is sought. The final control function will be suboptimal. However, with an approximate choice of functions S_i a good approximation to the optimal function can be obtained.

Each of the foregoing Newton–Raphson algorithms may employ any one of the techniques described in Section 8.3 to solve the subsidiary linear two-point boundary value problem. A procedure very similar to the successive sweep algorithm of Chapter 6 may be developed if the third algorithm is used in conjunction with the sweep method of 8.4.3.

8.6 INVARIANT IMBEDDING

In this section a somewhat different approach to the solution of two-point boundary value problems will be discussed. Invariant imbedding is the converse process of the method of characteristics, i.e., a two-point boundary value problem is converted to the problem of solving a partial differential equation instead of vice versa. Invariant imbedding is, conceptually, very similar to dynamic programming. As shall be shown, in certain circumstances the invariant imbedding equations may be replaced by a single dynamic programming boundary value problem.

Consider the two-point boundary value problem defined by

$$\dot{x} = f(x, p, t) \tag{8.77}$$

$$x(t_0) = x_0 \tag{8.78}$$

$$\dot{p} = h(x, p, t) \tag{8.79}$$

$$p(t_f) = \theta(x(t_f), t_f) \tag{8.80}$$

Clearly the solution to this problem can be determined if $p(t_0)$ is known. It is assumed that a differentiable function $P(x, t)$ exists such that

$$p(t) = P(x, t) \tag{8.81}$$

and hence $p(t_0) = P(x_0, t_0)$. Now an initial boundary value problem will be derived for $P(x, t)$. At the final time

$$P(x, t_f) = \theta(x, t_f) \tag{8.82}$$

Now differentiating Equation (8.81) with respect to time gives

$$\dot{P} = P_t + P_x \dot{x} \tag{8.83}$$

However $\dot{p}$ and $\dot{x}$ are given by $\dot{p} = h(x, P(x, t), t)$

$$\dot{x} = f(x, P(x, t), t) \tag{8.84}$$

and hence P satisfies the partial differential equation

$$P_t + P_x f(x, P, t) = h(x, P, t) \tag{8.85}$$

If the original boundary value problem is linear, this partial differential equation has a solution that is the same as that provided by the sweep method discussed in 8.4.3. The derivation of this solution is left to the reader as an exercise.

If the two-point boundary value problem was nonlinear, a Newton-Raphson algorithm may be developed. A nominal solution x^j, p^j is chosen which satisfies the differential equations (8.77), (8.79) and the terminal boundary condition equation (8.80) by integrating this equation backward in time with some arbitrary value for $x(t_f)$ completing the boundary conditions. Thus the only condition that will not be satisfied is the initial state condition equation (8.78). The algorithm must find a new nominal solution which more nearly satisfies the initial state.

Since the nominal solution satisfies Equation (8.85), neighboring solutions satisfy (at least to first order) the equation

$$\partial \dot{x} = (f_x + f_p P_x)\, \delta x \tag{8.86}$$

Now, choosing $\delta x(t_0)$ so that $\delta x(t_0) = x_0 - x^j(t_0)$ and integrating the equation (8.86) forward to the final time, a correction may be found for $x^j(t_f)$. Note, however, that this implies that $P_x(t)$ $t \epsilon [t_0, t_f]$ is available. The function P_x is obtained by differentiating Equations (8.82) and (8.85) with respect to x giving, as a set of terminal boundary conditions,

$$P_x(x_f^j, t_f) = \theta_x(x_f^j, t_f)$$

and as a set of n^2 characteristic equations

$$\dot{P}_x = -P_x[f_x + f_p P_x] + h_x + h_p P_x \tag{8.87}$$

It is left to the reader to verify that these equations are similar to those obtained from applying the sweep algorithm to the linearized equations

$$\delta\dot{x} = f_x\,\delta x + f_p\,\delta p$$
$$\delta\dot{p} = h_x\,\delta x + h_p\,\delta p$$

Unfortunately, if x and p are n-vectors, applying the invariant imbedding technique results in n partial differential equations. Conceptually, therefore, for optimization problems invariant imbedding is less desirable than dynamic programming which leads to only one partial differential equation. It is easy to show that the set of n partial differential equations for P for an optimal control problem are the same as those for V_x^{opt} which could be obtained by differentiating the partial differential equation for V^{opt}.

It is clearly of interest to examine the circumstances under which it is possible to convert the invariant imbedding equations into a single dynamic programming equation. A function V is sought such that

$$V_{x_i} = P_i, \qquad i = 1, \ldots, n \tag{8.88}$$

A necessary and sufficient condition for the existence of V is

$$\partial P_i/\partial x_i = \partial P_j/\partial x_i \tag{8.89}$$

Thus, at the final time this implies that the following equation holds

$$\partial \theta_i/\partial x_j = \partial \theta_j/\partial x_i \tag{8.90}$$

This is true if, and only if, there exists a function $\Phi(x)$ such that

$$\partial \Phi/\partial x_i = \theta_i(x) \tag{8.91}$$

It is also necessary to show that

$$\partial \dot{P}_i/\partial x_i = \partial \dot{P}_j/\partial x_i \tag{8.92}$$

Now $\dot{P}_x$ is given by Equation (8.87) and so $\dot{P}^i_{x_j}$ is

$$\dot{P}^i_{x_j} = -P^i_{x_k}[f^k_{x_j} + f^k_{p_n} P^n_{x_j}] + h^i_{x_j} + h^i_{p_k} P^k_{x_j} \tag{8.93}$$

and $\dot{P}^j_{x_i}$ is

$$\partial \dot{P}_j/\partial x_i = -P^j_{x_k}[f^k_{x_i} + f^k_{p_n} P^n_{x_i}] + h^j_{x_i} + h^j_{p_k} P^k_{x_i} \tag{8.94}$$

Thus Equation (8.92) holds if and only if,

$$h^i_{x_j} = h^j_{x_i}, \qquad f^i_{p_j} = f^j_{p_i}, \qquad f^i_{x_j} = -h^j_{p_i} \tag{8.95}$$

and these equations are satisfied if and only if there exists a function $H(x, p)$ such that

$$f^i = \partial H/\partial p_i , \qquad h^i = -\partial H/\partial x_i \tag{8.96}$$

Thus, if Equations (8.90)–(8.95) hold, there exists a function $V(x, t)$ such that Equation (8.88) is satisfied.

Next a boundary value problem for $V(x, t)$ will be derived. It is clear from Equation (8.91) that

$$V(x, t_f) = \Phi(x, t_f) \tag{8.97}$$

can be used as a terminal boundary condition for V. Now Equation (8.85) can be written in the form

$$V_{xt} + V_{xx}H_p = -H_x$$

or

$$\partial[V_t + H(x, V_x , t)]/\partial x = 0 \tag{8.98}$$

This equation will be satisfied if V is chosen to satisfy the partial differential equation

$$V_t + H(x, V_x , t) = 0 \tag{8.99}$$

Equations (8.97) and (8.99) define a boundary value problem for V. Of course, this problem is not unique as the constants of integration have been set equal to zero. This boundary value problem is the same as the one derived by dynamic programming.

PROBLEMS

1. Extend the maximum principle to problems with the following features: (a) control parameters, (b) control variable inequality constraints, (c) discontinuities.
2. State a Newton–Raphson algorithm not discussed in this chapter. What are the independent variables? What are the dependent variables? How is the nominal chosen? What type of Newton–Raphson method is it?
3. Develop the sweep algorithm for a linear two-point boundary value problem by means of invariant imbedding.
4. Derive the differential equation for P_x by applying the sweep algorithm to the linearized two-point boundary value problem.
5. Use the results in the last section to derive the dynamic programming equation from the maximum principle.
6. Show that Equation (8.95) implies Equation (8.96).

BIBLIOGRAPHY AND COMMENTS

One of the earlier papers indicating the applicability of classical variational theory to modern optimization problems is

Breakwell, J. V. (1959). "The Optimization of Trajectories," *J. Siam Control*, Vol. 7, pp. 215–247.

A source of the maximum principle that is often quoted is

Pontryagin, L. S., Boltyanski, V. G., Gamkrelidze, R. V., and Mischchenko, E. F. (1962). *The Mathematical Theory of Optimal Processes.* Wiley (Interscience), New York.

A very clear exposition of the maximum principle and its application is given by

Rozonoer, L. I. (1960). "Pontryagin's Maximum Principle in the Theory of Optimum Systems I, II and III," *Automation and Remote Control*, Vol. 20.

The close relationship between the maximum principle and results in the calculus of variations have been discussed by many authors. See, for example,

Hestenes, M. R. (1964). "Variational Theory and Optimal Control Theory," in *Computing Methods in Optimization Problems* (A. V. Balakrishnan and L. W. Neustadt, eds.). Academic Press, New York.

The derivation of the classical results by means of dynamic programming has been popularized by Dreyfus,

Dreyfus, S. E. (1965). *Dynamic Programming and the Calculus of Variations.* Academic Press, New York.

The introduction of the Newton–Raphson algorithm to problems in function space is credited to Kantarovich. See, for example,

Kantarovich, L. V. (1949). "On Newton's Method," *Trudy. Mat. Inst. Steklov*, 28, pp. 104–144.

Early applications of Newton–Raphson algorithms to problems in the calculus of variations are found in

Hestenes, M. R. (1949). "Numerical Methods Obtaining Solutions of Fixed End Point Problems in the Calculus of Variation," *Rand Report* RM–102.

Stein, J. L. (1953). "On Methods for Obtaining Solutions of Fixed End Point Problems in the Calculus of Variations," *J. Res. Natl. Bur. Std.*

The application of Newton–Raphson shooting methods to optimal control problems is given in

Breakwell, J. V., Speyer, J. L., and Bryson, A. E. (1963). "Optimization and Control of Non-Linear Systems Using the Second Variation," *J. Siam Control*, Ser. A, Vol. 1, p. 193.

Melbourne, W. G. and Sauer, C. G. (1962). "Optimum Thrust Programs for Power-Limited Propulsion Systems," *Astro. Acta.*, Vol. 8, p. 206.

Shooting methods are discussed in some detail in

Keller, H. B. (1968). *Numerical Methods for Two Point Boundary Value Problems.* Blaisdell, Boston, Massachusetts.

The concept of iterating about a nominal control function was suggested by

Kelley, H. J., Kopp, R. E., and Moyer, G. (1963). "A Trajectory Optimization Technique Based upon the Theory of the Second Variation," A.I.A.A. Astrodynamics Conference, Yale University.

The sweep generalization of this approach is given by

McReynolds, S. R. and Bryson, A. E. (1965). "A Successive Sweep Method for Solving Optimal Programming Problems." *Joint Autom. Control Conf.*, pp. 551–555, Troy, New York.

Mitter, S. K. (1966). "Successive Approximation Methods for the Solution of Optimal Control Problems," *Automatica*, Vol. 3. Pergamon Press, New York.

An algorithm which iterates about nominal $x(t)$ and $p(t)$ trajectories is described by

Moyer, H. G. and Pinkham, G. (1964). "Several Trajectory Optimization Techniques II," in *Computing Methods in Optimization Problems* (A. V. Balakrishnan and L. W. Neustadt, eds.). Academic Press, New York.

Quasilinearization methods have been applied to many problems by Bellman. See, for example,

Bellman, R. W. and Kalaba, R. E. (1965). *Quasilinearization and Non-Linear Boundary Value Problems.* American Elsevier, New York.

The technique of transforming a set of ordinary differential equations into a single partial differential equation was employed in classical mechanics, for example, in the Hamilton–Jacobi theory; see

Goldstein, H. (1959). *Classical Mechanics*, Chap. 8. Addison–Wesley, Cambridge, Massachusetts.

Another approach is given by

Gelfand, I. M. and Fomin, S. V. (1963). *Calculus of Variations*, Chap. 6. Prentice-Hall, Englewood Cliffs, New Jersey.

Although Ritz methods are very popular in obtaining approximate analytic solutions, few papers on numerical algorithms exploit the technique. One example is given by

Storey, C. and Rosenbrock, H. H. (1964). "On the Computation of the Optimal Temperature Profile in a Tubular Reactor Vessel," in *Computing Methods in Optimization Problems.* Academic Press, New York.

References to the stability of the matrix Riccati equation were given at the end of Chapter 4. Athans and Falb (1966) have other references,

Athans M. and Falb, P. (1966). *Optimal Control.* McGraw-Hill, New York.

For properties of the linearized Hamiltonian system, see

Broucke, R. A. and Lass, H. (1968). "Some Properties of the Solutions of the Variation Equations of a Dynamical System," *JPL Space Programs Summary, 37–50,* Vol. 3, p. 22.

APPENDIX

Conjugate Points

Under certain circumstances the second partial derivative of the return with respect to the state V_{xx}^{opt} may approach infinity. Points at which this phenomenon occur are referred to as *conjugate points*. Clearly, at a conjugate point the validity of the dynamic programming equations is questionable as the continuity of V_x^{opt} cannot be guaranteed. Thus, computationally it is quite impossible to extend the optimal solution past a conjugate point using the successive sweep algorithm.

However, solutions obtained via the maximum principle can be extended through a conjugate point to previous times. It is well known in the classical calculus of variations that such extended solutions are not optimal, and in this appendix a similar result is obtained from dynamic programming. Classically the proof of this result has followed from a formal mathematical treatment of the second variation. The dynamic programming approach outlined here is heuristic and does not employ the concept of the second variation.

First a few examples will be given to illustrate the nature of the results.

EXAMPLE 1. Consider the following example: Maximize the performance index J, where

$$J = \int_{-\pi/2}^{\pi/2} (x^2 - u^2)\, dt$$

where x is governed by the state equation $\dot{x} = u$ and $x(-\pi/2) = 0$.

The dynamic programming solution yields $u^{\text{opt}} = -V_{xx}^{\text{opt}} x$, where

$$\dot{V}_{xx}^{\text{opt}}(t) = -1 - (V_{xx}^{\text{opt}})^2, \qquad V_{xx}^{\text{opt}}(\pi/2) = 0$$

This equation may be integrated analytically and $V_{xx}^{\text{opt}}(t) = \tan(\pi/2 - t)$. Now, as $t \to 0$ from the right, $V_{xx} \to +\infty$, and, hence, a conjugate point exists at $t = 0$. Thus it would appear that the dynamic programming solution may *not* be extended to include the initial time or to anytime prior to the initial time.

However, we may easily verify that the solution $u = 0$ $(x(t) = 0)$ satisfies the maximum principle. The Hamiltonian H is given by $H = pu + x^2 - u^2$ where the adjoint variable p is given by $\dot{p} = -2x$ and $p(\pi/2) = 0$. The Hamiltonian is a maximum when $u = p/2$. Hence the two-point boundary problem becomes

$$\dot{x} = p/2, \qquad x(-\pi/2) = 0$$

$$\dot{p} = -2x, \qquad p(\pi/2) = 0$$

One solution is $p(t) = 0$, $x(t) = 0$, and, hence, $u(t) = 0$. Is this an optimal solution? The answer is *no*. Consider the following control, Figure A.1, which can be made arbitrarily close to the foregoing control $u = 0$

$$u^* = \begin{bmatrix} 0 & -\pi/2 \leqslant t < -\epsilon \\ c & -\epsilon \leqslant t \leqslant \delta \\ P(t)x & \delta \leqslant t \leqslant \pi/2 \end{bmatrix}$$

where $\dot{P} = -1 - P^2$ and $P(\pi/2) = 0$. Here ϵ and δ are positive quantities which can be made as small as is desired. The related value of the performance index J^* is

$$J^* = \int_{-\epsilon}^{\delta} \{(ct)^2 - c^2\}\, dt + c^2(\epsilon + \delta)^2\, P(\delta)$$

$$= c^2 \left[\left(\frac{\delta^3 + \epsilon^3}{3} \right) - (\epsilon + \delta) + (\epsilon + \delta)^2\, P(\delta) \right]$$

Now as $\delta \to 0$ it is clear that the term $(\epsilon + \delta)^2\, P(\delta)$ will dominate because $P(\delta) \to \infty$, and hence J^* will become positive, i.e., larger than the zero value obtained when using $u = 0$. Note that $| u^* |$ can be made arbitrarily small because c can be made arbitrarily small. Thus a solution neighboring the solution given by the maximum principle has been found which has a higher return. Hence, the solution given by the maximum principle is *not necessarily* optimal.

In order to gain some additional insight into the conjugate points in this particular example, a plot of the optimal trajectories for various values of the final state x_f is given in Figure A.2. An optimal trajectory satisfies $\dot{x} = V_{xx}^{\text{opt}}(t)\, x(t)$ and, hence, $x(t) = x_f \sin(t)$. One interesting

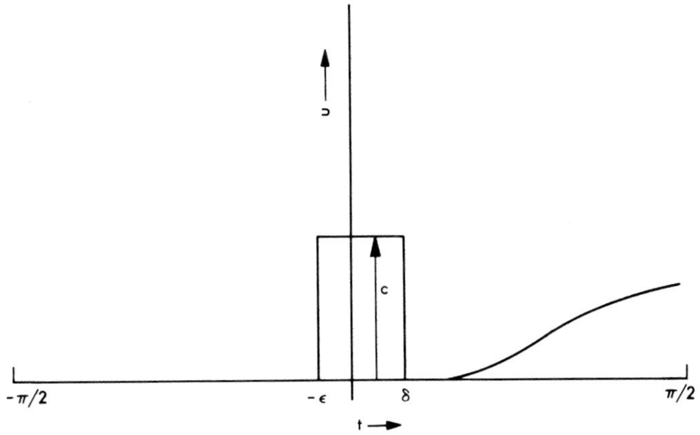

FIG. A.1. A control close to $u = 0$.

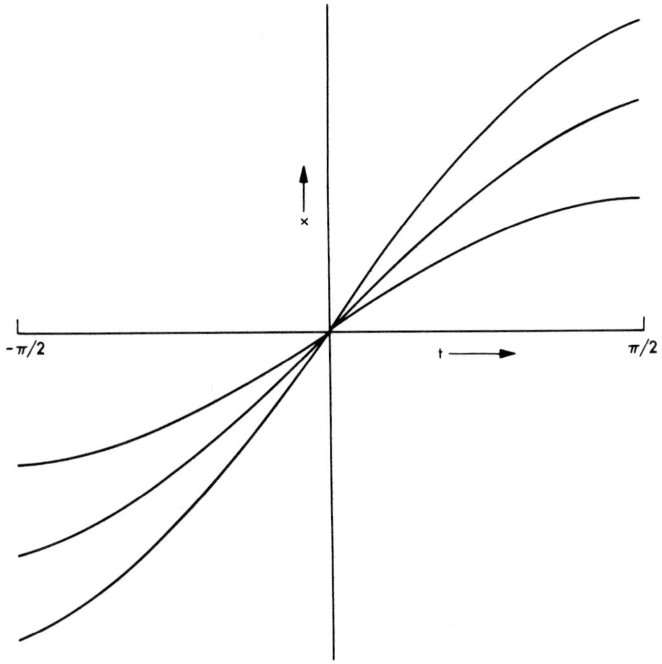

FIG. A.2. Optimal trajectories for various values of the final state.

property of these trajectories is that each one passes through the origin.

Also, it may be verified easily that each trajectory yields the same value of the performance index, namely zero. On the other hand, if at $t = 0$, x was nonzero, the maximum value of the performance index would be $+\infty$. Thus it is clear that if the initial time occurs before $t = 0$, it is possible to obtain a value of the performance index equal to $+\infty$. Hence, in this example it is not meaningful to seek an optimal solution for initial times prior to $t = 0$.

EXAMPLE 2. An example of a different type of phenomenon associated with conjugate points is illustrated by the following problem. Consider the problem of maximizing J where $J = \frac{1}{6}x^3(\pi/2) + \int_{-\pi/2}^{\pi/2} [x^2 - u^2] \, dt$, where x is governed by the state equation $\dot{x} = u$. This example is similar to Example 1 with the addition of a term in the performance index. The effect is that each trajectory now has a conjugate point at a *different time*. Along a trajectory V_{xx}^{opt} will satisfy the same differential equation. Thus, $V_{xx}^{\text{opt}}(t) = \tan(c - t)$ where c is determined from the terminal condition

$$V_{xx}^{\text{opt}}(\pi/2) = x(\pi/2) = \tan(c - \pi/2)$$

Clearly the locus of conjugate points are given by $t = c - \pi/2$. Taking the tangent of both sides gives $\tan(t) = \tan(c - \pi/2) = x(\pi/2)$. In order to find the locus of the conjugate points in terms of the state $x(t)$ and the time (t) the following relationship is required

$$x(t) = -x(\pi/2)^2/2 \cos t + x(\pi/2) \sin(t)$$

and hence $x(t) = \frac{1}{2} \tan(t) \sin(t)$. This locus of conjugate points forms an *envelope* (see Figure A.3). Each member of the field of optimal trajectories touches the envelope at a point, at which the tangents to the envelope and the trajectory coincide. Thus the rate of change of the return $\dot{V}$ is the same along the optimal path and the envelope at these points. This property leads to the following result: If $\tilde{u}(t)$ is the optimal control defined along the envelope and $\tilde{x}(t)$ is the corresponding trajectory

$$V^{\text{opt}}(\tilde{x}(t_0), t_0) = V^{\text{opt}}(x(t_1), t_1) + \int_{t_0}^{t_1} L(\tilde{x}(t), \tilde{u}(t), t) \, dt$$

This result is referred to as the *envelope* theorem. It states that the return for an optimal trajectory originating at a point on the envelope is the same even if the trajectory follows the envelope for an arbitrary distance before leaving the envelope.

As in the first example these optimal trajectories are not unique.

However, in this case a stronger result can be obtained. If the initial state lies on the envelope the dynamic programming solution is *not* optimal. This result was proved classically by combining the envelope theorem and the multiplier rule (maximum principle). The multiplier rule indicates that there exists a *unique* optimal trajectory if the Hamiltonian has a unique maximum. However, the envelope theorem demonstrates the existence of infinitely many trajectories with the same *optimal* value of the return, and hence these trajectories cannot be optimal. A dynamic programming proof of this result will not be given.

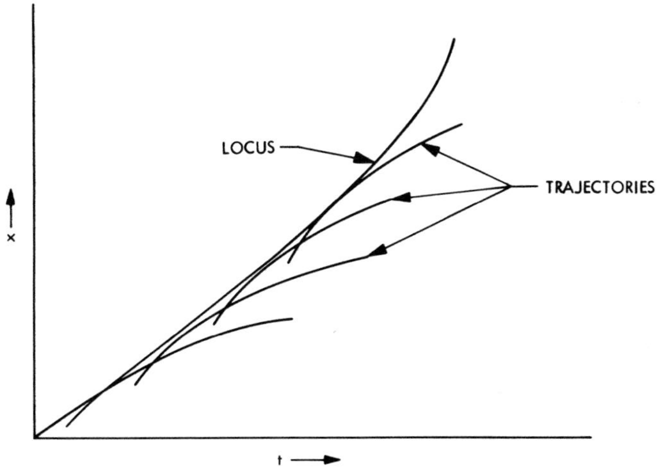

FIG. A.3. Locus of conjugate points.

EXAMPLE 3. *A General Result.* An interesting property of conjugate for maximization problems is that V_{xx}^{opt} will generally blow up to $+\infty$. This result is apparent from an inspection of the Riccati equation for V_{xx}^{opt}, viz,

$$\dot{V}_{xx}^{\text{opt}} = -L_{xx} - V_x^{\text{opt}} f_{xx} + (V_x^{\text{opt}} f_{xu} + L_{xu} + V_{xx}^{\text{opt}} f_u)(V_x^{\text{opt}} f_{uu} + L_{uu})^{-1}$$
$$\times (f_u^{\mathrm{T}} V_{xx}^{\text{opt}} + L_{ux} + f_{xu}^{\mathrm{T}} V_x^{\text{opt}})$$

Now assuming that f and L and their relevant partial derivatives are well behaved, V_{xx}^{opt} can become unbounded only in a finite time due to the last term. Since $V_x^{\text{opt}} f_{uu} + L_{uu} < 0$, this term will be negative.

Hence, when integrating the equation backward this term can only make $V_{xx}^{\text{opt}}(t)$ more positive and V_{xx}^{opt} must tend to $+\infty$ at conjugate points. It should also be noted that

$$\| V_{xx}^{\text{opt}} f_u \| \to +\infty \quad \text{at a conjugate point}$$

These results lead to the following theorem.

THEOREM 1. Let a solution to an optimal control problem be given that satisfies the maximum principle but that has an intermediate conjugate point. Then the solution is nonoptimal.

PROOF. The proof depends upon showing that a variation can be made positive and it is essentially based upon a generalization of the technique used in the first example. Consider the optimization problem without control parameters and assume that there is a conjugate point at time t_c where $t_0 \leqslant t_c < t_f$.

Now let δu^1 be an arbitrary control defined by

$$\delta u^1 = \begin{array}{ll} 0, & t_0 < t < t_c - \epsilon \\ c, & t_c - \epsilon < t < t_c + \delta \\ \delta u^*, & t_c + \delta < t < t_0 \end{array}$$

The variation of the performance index for this control becomes

$$\delta J = \int_{t_c-\epsilon}^{t_c+\delta} \{\delta x(L_{xx} + V_x)\,\delta x + 2\delta x(L_{xu} + V_u)c + c^{\text{T}}(L_{uu} + V_u)c\}\,dt$$
$$+ V_x\,\delta x(t_c + \delta) + \delta x(t_c + \delta)\,V_{xx}(t_c + \delta)\,\delta x(t_c + \delta) + \cdots$$

Now to first order in ϵ or δ and second order in c

$$\delta x(t_c + \delta) = f_u c(\epsilon + \delta) + \cdots + c^{\text{T}} f_{uu} c(\epsilon + \alpha)$$

Hence,

$$\delta J = c^{\text{T}}[(L_{uu} + V_x f_{uu})(\epsilon + \delta) + f_u{}^{\text{T}} V_{xx} f_u(\epsilon + \delta)^2]c \cdots$$

where $V_x^{\text{opt}} f_u = 0$ was employed.

Now since $f_u V_{xx} f_u$ tends to $+\infty$ as $\delta \to 0$, the second variation can be made positive and hence the solution is nonoptimal. Note that $\| \delta u \|$ can be as small as desired since c can be made arbitrarily small.

Index

Abnormality, 15, 90, 125
Accelerometer, control of, 99–101, 116
Aircraft landing problem, 133–140, 148, 165–170
Aircraft stabilizer, design of, 101–109, 116

Bang-bang control, problems, 183–184, 212
Boiler, optimal control of, 140, 148–149, 170–173, 182
Brachistochrone problem, 61–65, 71–73, 128–133, 162–165

Closed-loop solutions, 1, 8
Conjugate direction methods, 29, 32, 34
Conjugate points, 180, 235–240
Control parameters, 8, 36, 86, 122–123
Control variables, 36, 86
Convexity condition
  continuous optimal control problems, 179–180
  control problems with discontinuities, 204, 209–210
  discrete control problems, 45–46, 77–9
  parameter optimization problems, 10, 12–13

Direct methods, 26
Discrete maximum, principle, 67–68, 79, 84
Dynamic programming
  continuous optimal control problems, 87–114
  discrete optimal control problems, 37–45, 83
  numerical procedures, 54–56, 83

Finite difference methods, 63, 87–89, 115
First variation
  continuous optimal control problems, 144–148.
  discontinuities, 203
  discrete optimal control problems, 56–61, 65–67
  parameter optimization problems, 10
Free terminal time
  gradient method, 125–127
  successive sweep method, 159–162

Gradient method
  continuous optimal control problems, 117–149
  convergence properties, 17–19
  discontinuities, 187–188, 208
  discrete optimal control problems, 56–65, 83
  parameter optimization problems, 17–23
Gradient projection method, 33
  continuous optimization problems, 123–125
  discrete optimal control problems, 59–61
  parameter optimization problems, 19

Hamilton–Jacobi theory, *see* Dynamic programming

Indirect methods, 26
Invariant imbedding, 229–232, 234

Lagrange multipliers, 12–14, 41, 90
Linear quadratic problem
  continuous optimal control problem, 96–114, 116, 217–218
  discrete systems, 42–54, 83
  parameter optimization problem, 15
Local optimum, 9–14
Low-thrust transfer problem, 46, 73–77, 98, 113, 198–202

Maximum principle, 214–217

Neighboring extremal method,
  continuous optimal control problems, 174–177
  discontinuities, 190–191
  discrete optimal control problems, 79–81
Newton–Raphson method, *see* Successive sweep algorithm
  convergence properties, 25
  nonlinear two-point boundary value problems, 226–229, 233
  parameter optimization problems, 23–29

Open-loop solutions, 68
Optimal control law, 8, 39–41, 89, 91
Optimal return function, 8, 39, 89, 91

Penalty function technique, 19, 61
Performance index, 8, 36, 86
Principle of causality, 37–38
Principle of optimal feedback control, 39
Principle of optimality, 38

Quasilinearization, 228, 234

Return function, *see* Optimal return function, 39–40
Riccati equation, 98, 114, 115
  stability, 114, 223–226
Ritz methods, 229, 234

Second variation
  continuous optimal control problems, 177–179
  discontinuities, 204
  parameter optimization problems, 10
Shooting methods, 228, 233
State variables, 8, 35, 86
Stationarity condition
  continuous optimal control problems, 144–148
  discontinuities, 190–192, 209
  discrete optimal control problem, 65–67
  parameter optimization problems, 10, 12
Steepest ascent, *see* Gradient method, 122
Strengthened convexity condition
  discrete optimal control problem, 78
  parameter optimization problems, 11, 25
Successive sweep algorithm, *see* Newton–Raphson Algorithm, Sweep method
Successive sweep method
  continuous optimal control problems, 150–173, 181
  discontinuities, 188–190, 208–209
  discrete optimal control problems, 68–79, 84
Sufficiency theorem
  continuous optimal control problem, 179–180
  discrete optimal control problem, 77–79
  linear-quadratic discrete control problem 45–46
  parameter optimization problems, 11
Superposition, 218–219, 221–223
Sweep method, 220–226

Two-point boundary value problems, linear, 218–222

Vector notation, 4–6

Mathematics in Science and Engineering

A Series of Monographs and Textbooks

Edited by RICHARD BELLMAN, ***University of Southern California***

1. T. Y. Thomas. Concepts from Tensor Analysis and Differential Geometry. Second Edition. 1965

2. T. Y. Thomas. Plastic Flow and Fracture in Solids. 1961

3. R. Aris. The Optimal Design of Chemical Reactors: A Study in Dynamic Programming. 1961

4. J. LaSalle and S. Lefschetz. Stability by by Liapunov's Direct Method with Applications. 1961

5. G. Leitmann (ed.). Optimization Techniques: With Applications to Aerospace Systems. 1962

6. R. Bellman and K. L. Cooke. Differential-Difference Equations. 1963

7. F. A. Haight. Mathematical Theories of Traffic Flow. 1963

8. F. V. Atkinson. Discrete and Continuous Boundary Problems. 1964

9. A. Jeffrey and T. Taniuti. Non-Linear Wave Propagation: With Applications to Physics and Magnetohydrodynamics. 1964

10. J. T. Tou. Optimum Design of Digital Control Systems. 1963.

11. H. Flanders. Differential Forms: With Applications to the Physical Sciences. 1963

12. S. M. Roberts. Dynamic Programming in Chemical Engineering and Process Control. 1964

13. S. Lefschetz. Stability of Nonlinear Control Systems. 1965

14. D. N. Chorafas. Systems and Simulation. 1965

15. A. A. Pervozvanskii. Random Processes in Nonlinear Control Systems. 1965

16. M. C. Pease, III. Methods of Matrix Algebra. 1965

17. V. E. Benes. Mathematical Theory of Connecting Networks and Telephone Traffic. 1965

18. W. F. Ames. Nonlinear Partial Differential Equations in Engineering. 1965

19. J. Aczel. Lectures on Functional Equations and Their Applications. 1966

20. R. E. Murphy. Adaptive Processes in Economic Systems. 1965

21. S. E. Dreyfus. Dynamic Programming and the Calculus of Variations. 1965

22. A. A. Fel'dbaum. Optimal Control Systems. 1965

23. A. Halanay. Differential Equations: Stability, Oscillations, Time Lags. 1966

24. M. N. Oguztoreli. Time-Lag Control Systems. 1966

25. D. Sworder. Optimal Adaptive Control Systems. 1966

26. M. Ash. Optimal Shutdown Control of Nuclear Reactors. 1966

27. D. N. Chorafas. Control System Functions and Programming Approaches (In Two Volumes). 1966

28. N. P. Erugin. Linear Systems of Ordinary Differential Equations. 1966

29. S. Marcus. Algebraic Linguistics; Analytical Models. 1967

30. A. M. Liapunov. Stability of Motion. 1966

31. G. Leitmann (ed.). Topics in Optimization. 1967

32. M. Aoki. Optimization of Stochastic Systems. 1967

33. H. J. Kushner. Stochastic Stability and control. 1967

34. M. Urabe. Nonlinear Autonomous Oscillations. 1967

35. F. Calogero. Variable Phase Approach to Potential Scattering. 1967

36. A. Kaufmann. Graphs, Dynamic Programming, and Finite Games. 1967

37. A. Kaufmann and R. Cruon. Dynamic Programming: Sequential Scientific Management. 1967

38. J. H. Ahlberg, E. N. Nilson, and J. L. Walsh. The Theory of Splines and Their Applications. 1967

39. Y. Sawaragi, Y. Sunahara, and T. Nakamizo. Statistical Decision Theory in Adaptive Control Systems. 1967

40. R. Bellman. Introduction to the Mathematical Theory of Control Processes Volume I. 1967 (Volumes II and III in preparation)

41. E. S. Lee. Quasilinearization and Invariant Imbedding. 1968

42. W. Ames. Nonlinear Ordinary Differential Equations in Transport Processes. 1968

43. W. Miller, Jr. Lie Theory and Special Functions. 1968

44. P. B. Bailey, L. F. Shampine, and P. E. Waltman. Nonlinear Two Point Boundary Value Problems. 1968.

45. Iu. P. Petrov. Variational Methods in Optimum Control Theory. 1968

46. O. A. Ladyzhenskaya and N. N. Ural'tseva. Linear and Quasilinear Elliptic Equations. 1968

47. A. Kaufmann and R. Faure. Introduction to Operations Research. 1968

48. C. A. Swanson. Comparison and Oscillation Theory of Linear Differential Equations. 1968

49. R. Hermann. Differential Geometry and the Calculus of Variations. 1968

50. N. K. Jaiswal. Priority Queues. 1968

51. H. Nikaido. Convex Structures and Economic Theory. 1968

52. K. S. Fu. Sequential Methods in Pattern Recognition and Machine Learning. 1968

53. Y. L. Luke. The Special Functions and Their Approximations (In Two Volumes). 1969

54. R. P. Gilbert. Function Theoretic Methods in Partial Differential Equations. 1969

55. V. Lakshmikantham and S. Leela. Differential and Integral Inequalities (In Two Volumes). 1969

56. S. H. Hermes and J. P. LaSalle. Functional Analysis and Time Optimal Control. 1969.

57. M. Iri. Network Flow, Transportation, and Scheduling: Theory and Algorithms. 1969

58. A. Blaquiere, F. Gerard, and G. Leitmann. Quantitative and Qualitative Games. 1969

59. P. L. Falb and J. L. de Jong. Successive Approximation Methods in Control and Oscillation Theory. 1969

60. G. Rosen. Formulations of Classical and Quantum Dynamical Theory. 1969

61. R. Bellman. Methods of Nonlinear Analysis, Volume I. 1970

62. R. Bellman, K. L. Cooke, and J. A. Lockett. Algorithms, Graphs, and Computers. 1970

63. E. J. Beltrami. An Algorithmic Approach to Nonlinear Analysis and Optimization. 1970

64. A. H. Jazwinski. Stochastic Processes and Filtering Theory. 1970

65. P. Dyer and S. R. McReynolds. The Computation and Theory of Optimal Control, 1970

66. J. M. Mendel and K. S. Fu (eds.). Adaptive, Learning, and Pattern Recognition Systems: Theory and Applications, 1970

67. C. Derman. Finite State Markovian Decision Processes, 1970

68. M. Mesarovic, D. Macko, and Y. Takahara. Theory of Hierarchical Multilevel Systems, 1970

69. H. H. Happ. The Theory of Network Diakoptics, 1970

In preparation

G. A. Baker and J. L. Gammel. The Pade Approximant in Theoretical Physics

Karl Astrom. Introduction to Stochastic Control Theory

C. Berge. Principles of Combinatorics

Ya. Z. Tsypkin. Adaptation and Learning in Automatic Systems